KINGFISHER
POCKET BOOK OF
ASTRONOMY

James Muirden

Kingfisher Books

Editor: Vanessa Clarke
Illustrators: Ron Jobson,
Mike Saunders, Deborah
Mansfield

First published in 1982 as
Astronomy Handbook by
Kingfisher Books Limited
Elsley Court, 20-22
Great Titchfield Street
London W1P 7AD

This edition published 1983

BRITISH LIBRARY CATALOGUING
IN PUBLICATION DATA
Muirden, James
  [The Kingfisher astronomy
  handbook].
  Pocket book of astronomy.
  1. Astronomy — Observers' manuals
  I. Title   II. Pocket book of
  astronomy
  522   QB64
ISBN 0 86272 082 6

Printed in Italy by Vallardi
Industrie Grafiche, Milan

Contents

Introduction

Astronomy is a science, but it is also an exciting voyage of discovery. The sky is free for all to see, town-dwellers and country-dwellers alike. Everything, from the blinding Sun to the dimmest star, waits to be discovered.

Just as there are different kinds of object in the sky, so there are different types of astronomer. Some are referred to rather contemptuously by their more active cousins as 'armchair' astronomers as they get most of their enjoyment from reading and looking at pictures. But true amateur astronomy involves looking at the sky.

Amateurs and Professionals
Contrary to popular belief, the great majority of amateur astronomers do not possess a large telescope. Binoculars are the commonest astronomical instrument. Some enthusiasts do indeed have an

▼ **Comet Humason,**
1961. The stars have
trailed due to the camera
following the comet's
orbital motion.

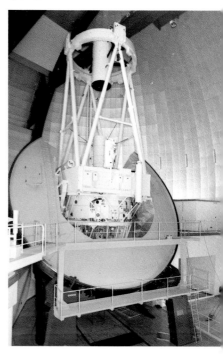

▼ **The 3·9-metre** Anglo-
Australian reflecting
telescope, New South
Wales: one of the modern
'giants'.

observatory in their gardens; but it is worth remembering that important discoveries have been made with the naked eye alone. After all, the ancient astronomers – in Greece, China and Egypt – gathered information with the simplest sighting instruments. Their method was to observe the sky and keep detailed records of everything they saw.

Professional astronomers today, of course, have the use of massive telescopes and other sophisticated equipment. For them the sky is literally the limit. Some amateurs go on to become professionals, but then a curious thing happens: they stop gazing upwards, and spend their time looking at dots on photographs or in the company of unromantic computers. So this book is for true amateurs, who want to go out beneath the sky and begin their voyage of astronomical discovery.

9

The Amateur Astronomer

Many amateur astronomers start out as star-gazers who regularly scan the night sky, making themselves familiar with the stars and their patterns. Some amateurs become particularly interested in the bodies that make up our solar system: the Sun, Moon, planets, comets and other smaller objects. Others prefer to look much further into space, examining the stars, star clusters and gaseous nebulae belonging to the Milky Way or Galaxy. They even search out other galaxies, so distant that light takes millions of years on its journey.

Amateur Discoveries

Important discoveries are sometimes made by amateurs, even in these days of giant telescopes. Some are made relatively near at hand; others out in remote regions of space. On Christmas evening 1980, for example, an English enthusiast, Roy Panther, discovered a new comet using a home-made telescope. Two months later, on February 24, 1981, an Australian amateur, the Reverend Robert Evans, was observing a remote galaxy of stars when he noticed a supernova, an exploding star, in its midst. Such an event is so rare that professional observatories immediately turned their telescopes on to it. The planets and the Moon have also revealed important secrets to amateur observers.

◀ **The night sky.**
Among these stars are some that orbit in pairs and others that change in brightness. From one night to the next, the sky is never the same.

10

▲ **Jupiter** — one of the best photographs taken from the Earth. The black spot is the shadow of one of its four major moons, and the Great Red Spot can be seen below it. Amateur observers have carried out valuable research on Jupiter.

▶ **The Moon's tumbled surface** has been thoroughly surveyed by spacecraft, but amateurs never tire of gazing at its mountains and craters.

Equipment – Binoculars

You can enjoy many hours of observation and see thousands of stars with no equipment at all, and this is how most amateurs start out. But binoculars and telescopes are useful. They allow you to see more objects in the sky and to examine them in more detail.

Binoculars

Binoculars consist of a pair of identical small telescopes with their light paths 'folded' to make them more compact. Although they are not as powerful as a true astronomical telescope, they have the great advantage of portability and comfortable viewing with both eyes.

An ordinary pair of binoculars will reveal about 30 stars for every single star seen with the naked eye. But they must be held steadily, which means resting them – or your elbows – on a solid support. Otherwise the stars will jump about, due to your heartbeat and muscular tension.

Binoculars do not provide very high magnification, so they will not reveal planetary details, although good ones will show three or four satellites of Jupiter and the crescent phase of Venus. They will also show the larger craters on the Moon, and through them sunspots can be projected on to a sheet of paper.

BINOCULARS

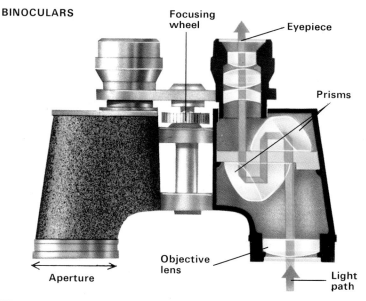

Focusing wheel

Eyepiece

Prisms

Objective lens

Aperture

Light path

▲ **A small hand telescope,** or binoculars, are an introduction to the sky for most people. Even the smallest instrument will show far more stars than the naked eye can see.

Aperture and Magnification

Binoculars and small hand telescopes belong to the same family of low-power instruments. The important features are the *aperture* and the *magnification*. The aperture is the diameter of the objective lens at the front of the instrument – usually between 30 and 50 millimetres. The larger this lens is, the brighter is the image of a star, since more light is being collected.

The magnification indicates how much larger an object looks. The Moon is half a degree across in the sky; through a × 10 telescope it will look five degrees across. The higher the magnification, the smaller is the amount of sky that can be seen at any one time.

Binoculars carry labels such as 8 × 30 and 10 × 50. In the first case, the magnification is × 8 and the objective is 30 millimetres across; the second has a magnification of × 10 and an aperture of 50 millimetres. For general astronomical work a 10 × 50 instrument is ideal.

13

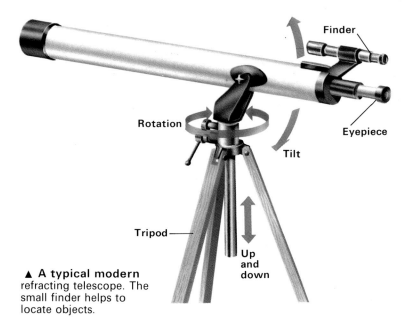

Finder

Rotation

Tilt

Eyepiece

Tripod

Up
and
down

▲ **A typical modern** refracting telescope. The small finder helps to locate objects.

◄ **The 1-metre refractor** of Yerkes Observatory, the world's largest.

▼ **Galileo's telescopes.** In 1610 he discovered Jupiter's moons.

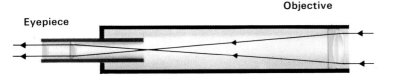

Eyepiece

Objective

Refracting Telescopes

Most telescopes are refractors: that is to say, they have an objective lens to collect and focus the light and form an image of the object near the other end of the tube (see the diagram above). This image is magnified by a small lens called the *eyepiece.* The objective lens should contain two glass components, one in front of the other; if a single lens is used, the image will be coloured and blurred. A two-lens objective is said to be *achromatic,* or colour-free.

The smallest useful refractor for astronomical purposes has an aperture of 60 millimetres, but 75 millimetres is much better, and anything larger is really powerful. A refractor's tube is usually about fifteen times as long as the aperture, and this is one disadvantage: refractors are not easy to mount rigidly. Many cheap refractors not only give poor definition, but are mounted on wobbly stands. A poorly-mounted telescope is useless, since no detail can be seen if the image vibrates.

BUYING A TELESCOPE

When buying any telescope, try to get an experienced observer to have a look through it before you finally decide. Get in touch with your local astronomical society about this (see page 183); they will be delighted to help. A poor telescope could turn you off astronomy before you have even begun! Most telescopes are supplied with several eyepieces. To find the magnification, divide the focal length of the eyepiece (which should be marked) into the focal length of the objective (the distance from the lens to the image it forms of a distant object). Three magnifications are best, for example:

Aperture	Magnifications		
60 mm	X 30	X 90	X 150
80 mm	X 40	X 120	X 200
100 mm	X 50	X 150	X 250

Low powers show more of the sky at one time. High powers are needed for fine details.

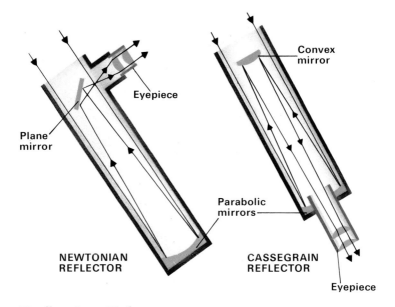

Convex mirror

Eyepiece

Plane mirror

Parabolic mirrors

NEWTONIAN REFLECTOR

CASSEGRAIN REFLECTOR

Eyepiece

Reflecting Telescopes

Reflecting telescopes are used only by astronomers. Their main attraction over refractors is that large mirrors are cheaper to produce than lenses of the same size. They are also more compact than refractors and can therefore be mounted more easily. The smallest commercially-available reflectors have a concave mirror about 100 millimetres across. But the best size for an amateur is 150 millimetres. A good telescope of this size can give a lifetime of enjoyment.

To give good definition, the mirrors in a reflecting telescope need to be polished accurately within about one tenth of the wavelength of visible light, or about 0·00005 millimetre of perfection. A Newtonian (see above) consists of a concave mirror of parabolic cross-section, and a small flat mirror to reflect the focused light through a hole in the side of the tube. A Cassegrain (also above) has a small convex mirror which sends the light back through the small hole cut in the main mirror.

Other more complicated reflectors have been developed but the Newtonian is the one most amateurs use. The Cassegrain type is much more expensive. Astronomical mirrors are usually coated with aluminium on the front surface. These coatings must be replaced occasionally and this is the main drawback of reflectors.

TELESCOPE MOUNTINGS

Telescopes are generally mounted in one of two ways. The simplest is the *altazimuth* mounting. This allows the tube to move vertically and horizontally. To follow a star you have to swivel the telescope up and around little by little and this can be awkward.

An *equatorial* mounting avoids this problem, allowing you to follow the path of a star with just one movement. It has two axes at right angles to each other like the altazimuth but one is parallel to the Earth's axis. If this (the polar axis) is turned once a day opposite to the direction in which the Earth spins, the telescope keeps pointing in the same direction (see the diagram on the right). The other axis – the declination axis – is used only when locating an object to begin with.

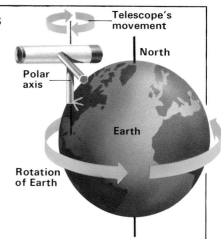

▲ **An equatorial mounting** has one axis parallel to the Earth's.

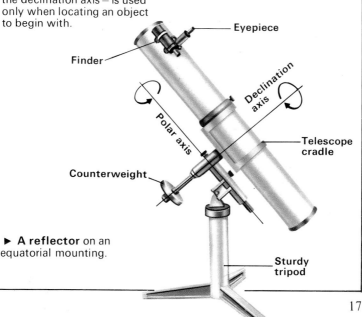

▶ **A reflector** on an equatorial mounting.

17

▼ **A 210-millimetre reflector,** professionally made and on a sturdy equatorial mounting.

▲ **A home-made 215-millimetre altazimuth** reflector made almost entirely of wood and used by the author.

▼ **The dome** of one of the world's largest telescopes: the 4-metre Mayall reflector at Kitt Peak Arizona.

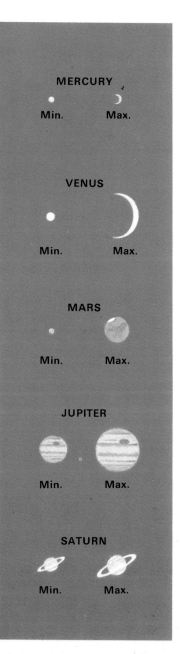

MERCURY

Min.   Max.

VENUS

Min.   Max.

MARS

Min.   Max.

JUPITER

Min.   Max.

SATURN

Min.   Max.

Almost all professional instruments are reflectors; the last big refractor, the one-metre telescope at Yerkes, USA, was built in 1897.

Many amateurs have built their own reflectors, using purchased optics and making the tube and mounting. The 215-millimetre reflector pictured opposite was made almost entirely of wood. It is on a simple *altazimuth* mounting.

Magnification

The illustration on the left gives you an idea of what an amateur can expect to see. If you hold the book about 25 centimetres away from you, it shows the planets as seen through a telescope magnifying 200 times. Each planet is shown near its maximum and minimum possible distance from the Earth. (Uranus, Neptune and Pluto are omitted).

But the telescopic view is never as steady as this; currents in the atmosphere make the image flicker and blur. And a small telescope will reveal less detail than a large one.

A Telescope's View

Most astronomical telescopes give an upside-down view. Terrestrial telescopes use extra lenses to give an upright image. Astronomers do without these because any piece of glass in the light beam makes the image fainter.

The Moving Sky

The base from which we observe the sky is moving all the time – but we are so used to living on the spinning Earth that we assume it is stationary, and that everything in the universe moves around it. We seem, like the observer in the diagram below, to be at the centre of a huge rotating invisible sphere – the celestial sphere. The heavenly bodies seem to be attached to the inside of this sphere and carried around it in an east to west direction.

Since the Earth revolves on its north-south axis, the celestial sphere appears to rotate around the same axis. If a camera is directed towards one of the celestial poles, and the shutter is left open for a few minutes while the Earth spins, the stars form trails centred on the pole.

Ptolemy's Universe
The illusion that the stars, planets and Sun revolve around the Earth is so convincing that people believed in it for thousands of years. The Greek astronomer Ptolemy (about AD 140) worked out a

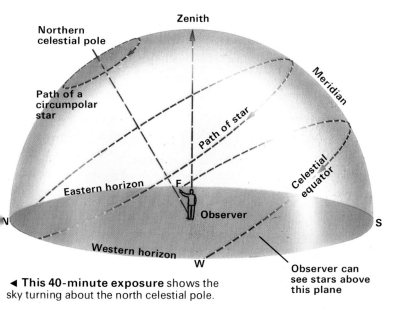

◀ **This 40-minute exposure** shows the sky turning about the north celestial pole.

Observer can see stars above this plane

21

complicated system in which the different planets, the Sun, Moon and stars, all had their own invisible shells or spheres rotating around the Earth. These separate spheres were necessary because only the stars keep the same relative positions from night to night. The planets and the Moon wander across the sky.

The Spinning Earth
Today we know that the Earth and everything we see is moving. The stars appear in the same place but they are not stationary. Some are moving very fast, but they are so far away that their position in relation to each other does not seem to change. The planets are nearer and their movement more noticeable. They revolve around the Sun and so their positions on the celestial sphere slowly change. The Moon takes only a month to go round the celestial sphere once. The Sun takes a year – the length of time taken by the Earth to complete one orbit.

▼ **The planets** appear to move around the celestial sphere in an irregular manner, simply because the Earth also moves. In Ptolemy's Earth-centred universe, these irregularities had to be explained by small epicycles, as shown here.

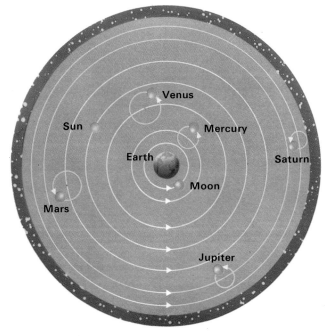

PROVING THE EARTH'S ROTATION

Foucault's famous experiment to prove that the Earth rotates was carried out in Paris in 1851. A heavy weight tends to keep swinging in the same direction while the Earth turns underneath it, making the pendulum's direction appear to change.

You can carry out this experiment for yourself with a long thread and a weight. The longer the thread, the better, and the weight should be as heavy as possible. A good place to hang the thread is at the top of a staircase but draughts, vibrations and twists in the thread must be avoided if the pendulum is to keep swinging in the same plane. Mark the path of the pendulum at the start on a simple grid and return later to observe the apparent change of direction.

Thread (at least 6 metres long)

▲ **A Foucault pendulum** at the Science Museum, London. The pendulum is almost 25 metres long and the bob (weight) over $13\frac{1}{2}$ kilogrammes.

Weight (at least 5 kilogrammes)

A

B

THE APPARENT PATH OF ORION

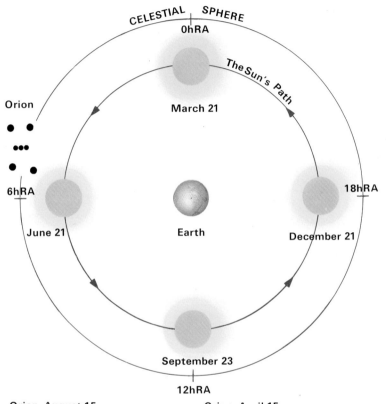

CELESTIAL SPHERE

0hRA

The Sun's Path

Orion

March 21

6hRA

18hRA

June 21

Earth

December 21

September 23

12hRA

Orion August 15

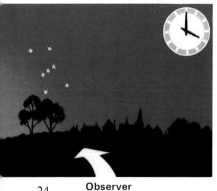

Observer

Orion April 15

Observer

Constellations and the Seasons

The Earth's annual journey around the Sun makes our star appear to travel round the celestial sphere once in a year. This affects the visibility of the constellations, since stars cannot be seen in the daytime.

Look, for example, at the diagram on the left. It shows the apparent movement of the Sun as the Earth orbits, and the famous constellation of Orion. In June the Sun lies in the direction of Orion, which is therefore in the daytime sky and invisible. By August, however, the Earth's orbital motion has carried the Sun some way east of Orion and it now rises in the morning sky before dawn. By midwinter, the Sun appears opposite Orion in the sky, and the constellation is well placed for midnight viewing. By April, the Sun has moved in close on its western side, and Orion disappears into the evening twilight once more.

This is the same for almost all the constellations. Because of the Earth's movement around the Sun, each is visible at different times of the year – some only in winter, others only in summer.

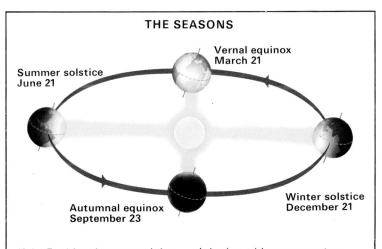

THE SEASONS

Vernal equinox
March 21

Summer solstice
June 21

Autumnal equinox
September 23

Winter solstice
December 21

If the Earth's axis were upright, the Sun would pass directly over the equator every day. But the axis is tilted in a fixed direction, $23\frac{1}{2}°$ from the vertical.

On June 21, the north pole is inclined towards the Sun, bringing midsummer to the northern hemisphere, while the south experiences midwinter. On December 21 the situation is reversed, while spring and autumn occur between these times.

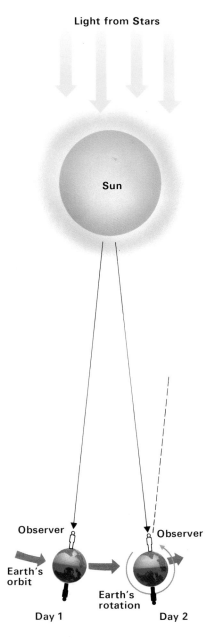

Light from Stars

Sun

Observer

Earth's
orbit

Earth's
rotation

Observer

Day 1

Day 2

Solar Days and Sidereal Days

Since life on Earth is regulated by day and night, the Sun is used as the basic timekeeper. The normal Sun or *solar* day corresponds to the interval between successive noons or midnights – that is the 24 hours in which the Earth spins once on its axis.

But as the Earth orbits the Sun at the same time as it spins, the solar day does not indicate the true rotation of the Earth in space with respect to the stars.

Look at the diagram on the left. On Day 1 it is noon to one observer and midnight to the other. One *sidereal* day later (Day 2), it is not quite noon and midnight again, since the Earth has moved a little way along its orbit, and must spin slightly more to bring the Sun back to where it was.

Since most astronomers observe the Moon, planets and stars, their telescopes are adjusted to turn on their polar axis in one sidereal day (23h 56m), the time the celestial sphere takes to rotate once. The amateur must remember that each month the constellations will be in the same positions two hours earlier and that every night the stars appear to rise four minutes earlier. On page 68, you will find how Sidereal Time indicates which constellations are well placed for observation.

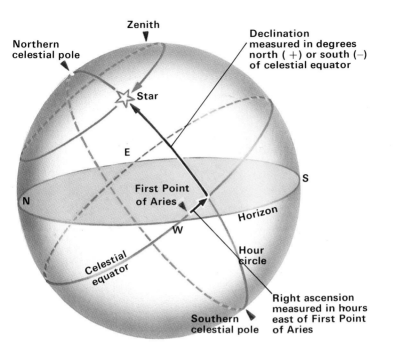

CELESTIAL CO-ORDINATES

The position of a city or river on Earth can be found on a map using latitude and longitude The position of an object on the celestial sphere is also described in these terms but latitude is called *Declination* or *Dec* and longitude *Right Ascension* or *RA*. These are the celestial or astronomical co-ordinates.

Declination is reckoned in degrees north (positive) or south (negative) of the celestial equator

– the line dividing the celestial sphere into two halves, in the plane of the Earth's equator.

Right Ascension is divided into 24 sidereal hours, measured eastwards from 0h, which is the RA of the Sun on the first day of northern spring (March 21).

The Sidereal Time is equal to the RA that is due south (or due north to a southern observer) at any instant, and is shown by special astronomical clocks.

Constellations and the Sun

The stars are fixed on the celestial sphere, unlike the Sun, Moon and planets. The Sun follows a path which is called the *ecliptic*. This is the projection, on the celestial sphere, of the plane of the Earth's orbit around the Sun. The ecliptic crosses the celestial equator at an angle equal to the Earth's axial tilt of $23\frac{1}{2}$ degrees.

The orbital planes of the Moon and planets coincide fairly closely with that of the Earth. They are always to be found within a few degrees of the ecliptic, in a band known as the *Zodiac*. The word means 'circle of animals' and refers to the constellations: Aries, Taurus, Gemini, Cancer, Leo, Virgo, Libra, Scorpius, Sagittarius, Capricornus, Aquarius and Pisces. These star groups make up the sequence of constellations through which the Sun, Moon and planets all pass.

▼ **This map shows** the Sun's path or ecliptic around the celestial sphere. Its position on the first day of each month is indicated by a yellow circle. The planets are always to be found within the 18° band, the Zodiac, centred on the ecliptic.

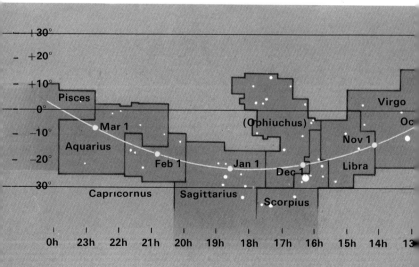

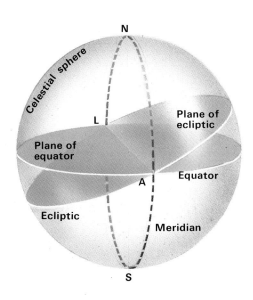

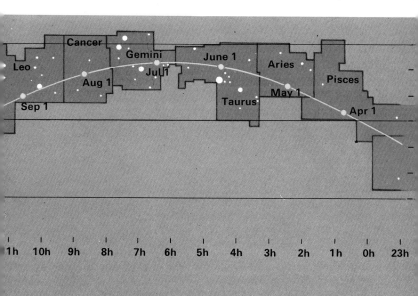

The Sun – Our Star

There is nothing strange about the Sun, apart from its closeness. It is about a quarter of a million times nearer to us than the next known star, which means that we can examine it in far more detail than any of its neighbours. Solar astronomy is a particularly important subject, and amateur observers can enjoy studying the Sun's surface provided that they take proper precautions (see pages 36 and 37). **Remember that you must never look directly at the Sun.**

Long before serious astronomy began, the Sun's daily motion across the sky and annual movement around the celestial sphere were studied. Farmers, for example, would use the Sun's position to tell them when they should plant crops. As has been shown on page 25, the seasons are caused by the Earth being tilted on its axis. This also affects the position of the Sun around the year. In winter it never rises as high in the sky as it does in summer, and the points on the horizon at which it rises and sets shift with the seasons.

▼ **The Sun's daily path** across the sky changes with the seasons of the year.

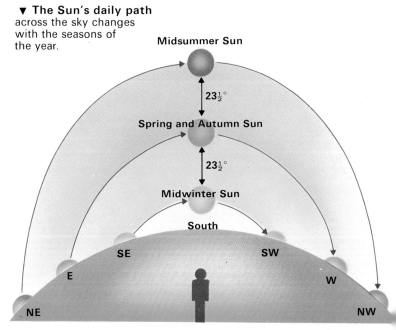

Midsummer Sun

$23\frac{1}{2}°$

Spring and Autumn Sun

$23\frac{1}{2}°$

Midwinter Sun

South

SE   SW

E   W

NE   NW

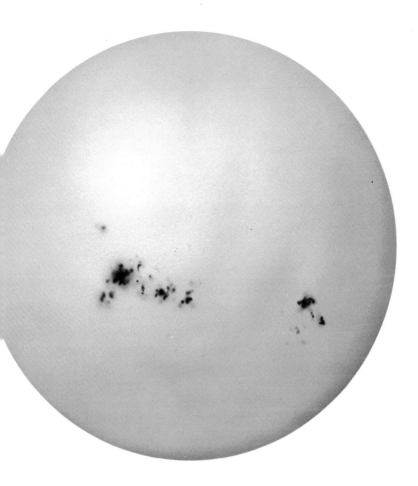

SUN FACTS

Diameter: 1,392,000 kilometres (109 X Earth)
Mass: 328,900 X Earth
Volume: 1,300,000 X Earth
Surface temperature: 6000°C
Core temperature: about 15,000,000°C
True rotation period: 25·38 days
Apparent rotation period: 27·28 days
Mean distance from Earth: 149,600,000 kilometres
Cosmic year (time to orbit Galaxy): 225 million years
Estimated age: 4600 million years

31

◄ **In most sundials,** the gnomon, which casts the shadow, is parallel to the Earth's axis. The plate, on which the shadow falls and which carries the hour lines, can be fixed to a vertical wall, or set horizontal, as in the common sundial seen here.

Time and the Sun

The shadow cast by a sundial indicates the approximate time of day. It is rarely exactly right, because the Sun wanders several degrees east and west of its 'true' position.

This happens mainly because the Earth's orbit is slightly elliptical, and its orbital velocity varies, making the Sun appear to move fast or slow. Clocks are therefore regulated to Mean Solar Time, known as Greenwich Mean Time (GMT) or Universal Time (UT). The

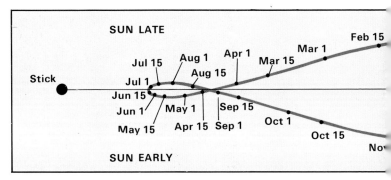

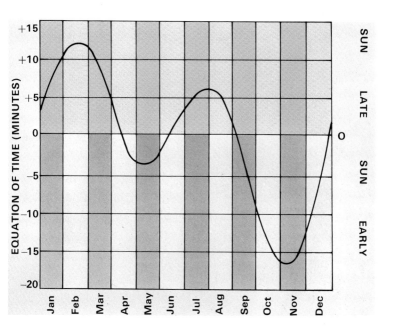

difference between this and sundial time (or Apparent Solar time) is known as the *equation of time* (see above).

Simple Sundials

The simplest sundial consists of an upright stick in the ground.

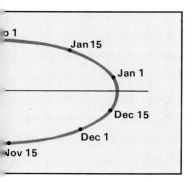

When the shadow points exactly north (or south, in the southern hemisphere), it is Apparent Noon. The equation of time correction will give True Noon, but this will only agree with Mean Solar Time if the sundial is exactly on one of the world's standard meridians.

In summer, the shadow cast by the stick is shorter than it is in winter. If the position of the shadow's tip is marked at True Noon every few days throughout the year, a shape like that shown below left, the *analemma,* will be produced.

Instructions for making and using a more sophisticated sundial will be found on page 168.

Sunspots and the Solar Cycle

The Sun has shone steadily for thousands of millions of years, but its surface, the *photosphere,* does not always look the same. Dark spots come and go and the number visible follows a cycle of about 11 years. This cycle also affects the shape of the Sun's faint atmosphere, or *corona,* which can be seen only during a total eclipse.

A sunspot is caused by a very strong magnetic field generated in the swirling material beneath the photosphere. Radiation from the interior is sucked away, leaving a cooler, darker area above. Sunspot interiors are about a thousand degrees cooler than the photosphere, but they are still hotter than the surfaces of many stars, and appear dark only by contrast.

If the image of a large sunspot is projected on to a screen, it is seen to consist of a dark centre (the *umbra*) and a lighter surrounding area (the *penumbra*). Many spots occur in pairs, and a large group will contain dozens of separate specks. A well-developed group lasts for several weeks or even months, and may be ten times the Earth's diameter in extent. At maximum activity, a dozen different groups may be visible at the same time and, as the Sun rotates, new sunspots come into view.

▼ **Sunspot groups** may be several times as large as the Earth.

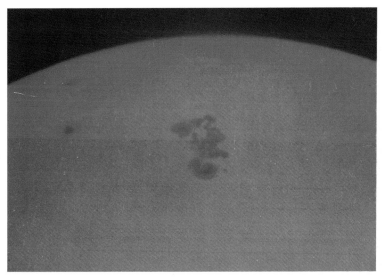

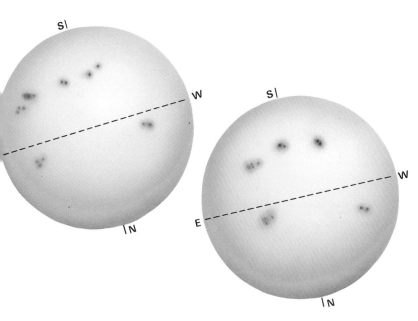

▲ These drawings of the Sun were made by projecting its image with binoculars at an interval of two days. The dotted line represents its equator.

▼ The Sun being viewed by projection, using binoculars. A simple wooden stand helps to keep the image sharp and steady.

35

Observing the Sun

It is highly dangerous to look at the Sun with the naked eye and you must never look at the Sun directly through telescopes or binoculars. People have been blinded in this way. The Sun's awesome radiation will destroy the eye's sensitive nerves in seconds.

Fortunately, there is a perfectly safe and sensible way of observing sunspots: by projecting the solar image on to a sheet of white paper. This can be done using an ordinary telescope (refractor or reflector), or a pair of binoculars.

The white paper is held some way behind the eyepiece (about 30 centimetres will do for a trial), and the eyepiece is adjusted until the round image of the Sun is sharp. The size of the solar image can be increased by moving the screen further back from the eyepiece.

The image must be screened from direct light, or its details will be washed out. A projection box (right), with a hole cut in the side to reveal the image, gives a very bright view. But even a simple shade will be adequate. A viewing box is particularly useful if a highly-magnified image is being projected, because it improves the contrast of the much fainter image.

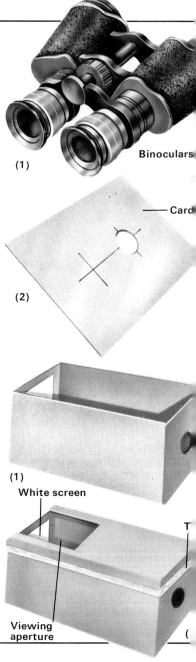

Binoculars

(1)

Card

(2)

(1)

White screen

T

Viewing aperture

(

A pair of binoculars like these can reveal plenty of sunspots (1).

To make a shade, draw the outline of the lenses on a piece of card and cut out one of the shapes as shown (2).

Secure the card with tape (3).

Prop the binoculars on a chair by the window and focus the Sun's image on to a white screen (4).

Curtains

Shade

(4)

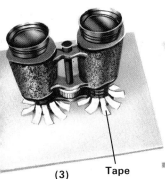

(3) **Tape**

White paper

For solar projection with a telescope a shoebox works well. Cut a hole in one end, to take the tube (1).

Stick a white screen at the other end, and cut out a viewing aperture (2).

Fit the box to the tube, and focus the image (3).

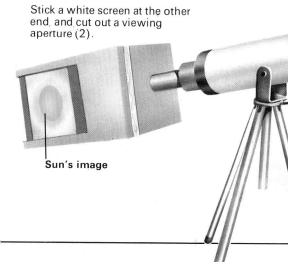

(3)

Sun's image

37

Observation Notes

Observing sunspots is, or should be, a daily business. The simplest method is to count the groups that are visible. After each month, add up the total number of groups observed and divide this by the number of days on which observations were made. The result is the Mean Daily Frequency (MDF) for the month – see below right.

Another interesting project is to draw the solar disc with its spots. The easiest way is to make a projection grid (opposite) and focus the Sun's image on to it.

Note the positions of the spots on the grid. Then copy them on to a sheet of thin paper, bearing the Sun's circular outline, laid over an identical grid so that the lines show through.

Before starting, however, you must fix the cardinal points on the disc. Leave the telescope fixed, and let the Sun's image drift across the screen. Twist the grid so that a sunspot trails accurately along the east-west line, and you know that the image is orientated correctly. Most astronomical telescopes give an inverted view.

▼ **A reflecting telescope** used to project the Sun's image. One advantage of this method is that several people can view the spots together.

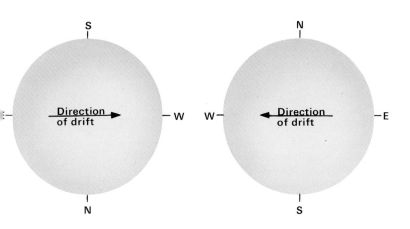

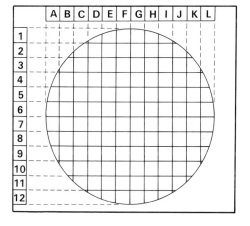

▲ **The orientation** of the projected image, using binoculars and an inverting astronomical telescope respectively (southern observers should reverse these directions).

◄ **A projection grid** for sunspot positions.

▼ **Sunspot activity** as recorded by the author using a 90-millimetre refractor.

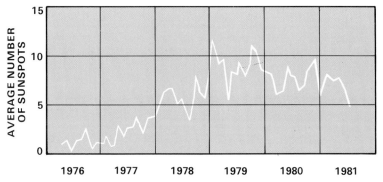

Solar Eclipses

▶ A total solar eclipse reveals the Sun's corona.

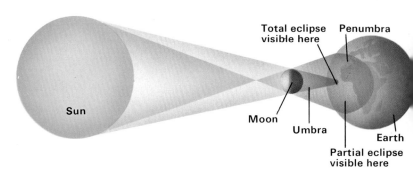

By a strange coincidence, the Sun and the Moon appear about the same size in the sky. When the Moon happens to pass in front of the Sun, it can block out the solar disc completely. Darkness falls, and the faint outer atmosphere or corona shines out around the black lunar outline during the seconds or minutes that the total eclipse lasts.

To see a total eclipse, you must be within the Moon's shadow as it sweeps over the Earth's surface, and this shadow is rarely more than a few hundred kilometres wide. Outside the central shadow, or umbra, is the wide penumbra, in which only a partial eclipse is seen.

In addition to the pale corona, the red prominences – colossal eruptions of hot gas, many times the size of the Earth – can often be seen shining around the edge of the Moon. Such a sight is well worth a long journey!

TOTAL SOLAR ECLIPSE TABLE		
Date	Maximum duration	Area of Visibility
1981 Jul 31	2m 03s	USSR, North Pacific Ocean
1983 Jun 11	5m 11s	Indian Ocean, East Indies, Pacific Ocean
1984 Nov 22-23	1m 59s	East Indies, South Pacific Ocean
1985 Nov 12	1m 55s	South Pacific Ocean, Antarctica
1987 Mar 29	0m 56s	Atlantic Ocean
1988 Mar 18	3m 46s	Indian Ocean, East Indies, Pacific Ocean
1990 Jul 22	2m 33s	Finland, USSR, Pacific Ocean

40

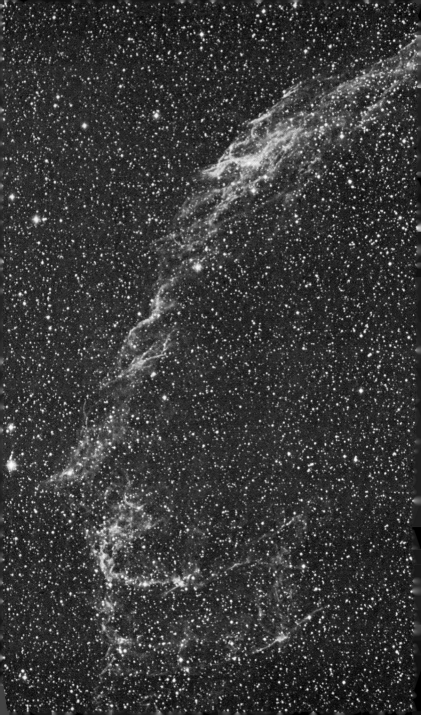

The Stars

When the stars shine out on a clear night, the sight is both confusing and awe-inspiring. Any attempt to organize them into constellations appears hopeless. The only definite first impression you may have is that some stars are very bright, while others can only be glimpsed.

In fact, this simple observation raises an important issue. Suppose that the stars have the same true brightness or luminosity. Then the faint ones must be more remote than the bright ones, just as a nearby street-lamp outshines one farther down the road. But if instead you suppose that the stars are equally distant, then the bright-looking ones must really be more luminous than the faint ones.

The early astronomers, who believed that the stars were attached to an invisible sphere, clearly held the second opinion. But one of the first men to study the stars seriously, William Herschel (who in 1781 discovered the planet Uranus), worked on the assumption that the stars are all equally luminous, and tried to gauge their relative distances by measuring their brightness in the sky.

Neither simple theory is correct. Some stars are over a million times more luminous than others, while the closest ones are thousands of times nearer than the most remote that have been detected in the Galaxy.

Brightness and Magnitude

The brightness of a star is measured in *magnitudes*. One magnitude step means a brightness ratio of 2·512. This number is chosen so that a 5-magnitude difference corresponds to a brightness ratio of exactly 100. In other words a 6th magnitude star is exactly 100 times fainter than a 1st magnitude star. The larger the number, the fainter the star. Stars that are brighter than magnitude 0 stars are given a negative magnitude.

Apparent magnitude indicates brightness in the sky. The faintest naked-eye stars are about magnitude 6 (often written mag 6). The brightest star in the sky, Sirius, is magnitude –1·47. The faintest

◀ **The Veil Nebula** in Cygnus, an enormous 'bubble' of material from an ancient star explosion. It now measures about five times the Moon's diameter in the sky.

stars detectable with a 150-millimetre telescope are about magnitude 13, or over 600,000 times fainter than Sirius.

Absolute magnitude indicates a star's luminosity; it is the apparent magnitude it would have if viewed from a distance of 32·6 light-years. (A light-year (l.y.) is the distance light travels in one year: about nine and a half million million kilometres.) The most luminous known stars are about magnitude –7. The Sun's absolute magnitude is 4·8 – about 40,000 times fainter. However, some nearby dim stars of about a millionth of the Sun's luminosity have been detected, so that it does lie in the upper half of the brightness table.

Distances in the Sky

Distances between the stars are so great that remote Pluto seems only a step away. If the Sun is represented by a table-tennis ball, Pluto will be a speck of dust about 150 metres away, while the nearest star will be another ball about 1000 kilometres away. The compactness of the solar system is difficult to visualize.

▼ **The patterns** that the stars make in the sky do not reveal their distribution in space. The stars forming the constellation of Orion appear to be projected on the celestial sphere in the pattern shown below left. In fact they are at very different distances from the Sun as shown in the illustration below right.

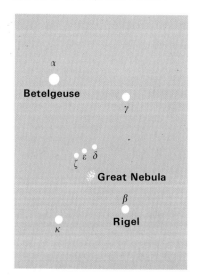

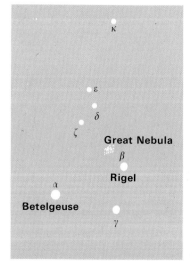

THE BRIGHTEST STARS IN THE SKY

Star	Constellation	Magnitude App.	Magnitude Abs.	Type	Distance (l.y.)
Sirius (α)	Canis Major	−1·47	1·4	Dwarf	8·7
Canopus (α)	Carina	−0·7	−3·3	Supergiant	110
Rigel Kent (α)	Centaurus	−0·3	4·4	Dwarf	4·3
Arcturus (α)	Bootes	−0·1	−0·3	Giant	36
Vega (α)	Lyra	0·0	0·6	Dwarf	26
Rigel (β)	Orion	0·1	−7·0	Supergiant	850
Capella (α)	Auriga	0·1	−0·6	Giant	45
Procyon (α)	Canis Minor	0·3	2·6	Subgiant	11
Achernar (α)	Eridanus	0·5	−1·6	Subgiant	75
Hadai (β)	Centaurus	0·6	−4·4	Giant	330
Altair (α)	Aquila	0·8	1·9	Giant	16
Acrux (α)	Crux	0·8	−3·9	Subgiant	280
Betelgeuse (α)	Orion (var)	0·8	−5·5	Supergiant	650
Aldebaran (α)	Taurus	0·9	−0·3	Giant	65
Spica (α)	Virgo	1·0	−3·5	Dwarf	260
Antares (α)	Scorpius (var)	1·1	−4·5	Supergiant	430
Pollux (β)	Gemini	1·15	0·2	Giant	35
Fomalhaut (α)	Piscis Austrinus	1·2	1·7	Dwarf	23
Mimosa (β)	Crux	1·2	−5·0	Giant	570
Deneb (α)	Cygnus	1·3	−7·0	Supergiant	1500

STAR SIZES

Star sizes vary greatly. Dying white dwarf stars (page 49) are too small to be shown here. The Sun is much smaller than giant and supergiant stars like Antares and Betelgeuse, but these are huge because they have been puffed up by internal pressure. Their average density is about a thousandth of the air we breathe.

It is important to remember that star masses (the amount of material in them) do not vary by nearly as much as do their diameters. Even a large star like Betelgeuse, whose volume is millions of times greater than the Sun's, has only about twenty times its mass.

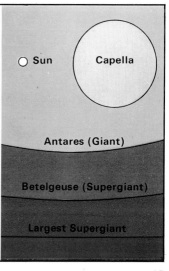

Sun Capella

Antares (Giant)

Betelgeuse (Supergiant)

Largest Supergiant

The brightness of a star in the sky is no sure guide to distance, since star luminosities vary so much. But some nearby stars give astronomers a clue: careful measurements show that they are slowly changing their position with respect to the star patterns. All the stars in the Galaxy are shooting through space at speeds of many kilometres per second, and the nearer they are the more noticeable this motion becomes.

Known as *proper motion,* this movement of the stars is the clue to likely neighbours. Even the largest proper motion, however, amounts to a shift equal to the Moon's width in 180 years, and most are only a tiny fraction of this.

THE NEAREST STARS

Star		Constellation	Magnitude App.	Abs.	Type	Distance (l.y.)	Proper motion (″/century)
Proxima		Centaurus	10·7	15·1	Dwarf	4·3	387
Rigel	A	Centaurus	0·0	4·4	Dwarf	4·3	367
Kent (α)	B	Centaurus	1·4	5·8	Dwarf	4·3	367
Barnard's star		Ophiuchus	9·5	13·2	Dwarf	5·2	1030
Wolf 359		Leo	13·5	16·7	Dwarf	7·6	467
Lalande 21185		Ursa Major	7·5	10·5	Dwarf	8·1	477
Sirius	A	Canis Major	−1·5	1·4	Dwarf	8·7	307
(α)	B	Canis Major	8·5	11·4	White dwarf	8·7	307
UV	A	Cetus	12·5	15·3	Dwarf	8·9	336
	B	Cetus	13·0	15·8	Dwarf	8·9	336
Ross 154		Sagittarius	10·6	13·3	Dwarf	9·5	74
Ross 248		Andromeda	12·2	14·7	Dwarf	10·3	182
ε		Eridanus	3·7	6·1	Dwarf	10·7	98
Ross 128		Virgo	11·1	13·5	Dwarf	10·8	136
L 789–6		Aquarius	12·2	14·6	Dwarf	10·8	327
61	A	Cygnus	5·2	7·5	Dwarf	11·2	520
	B	Cygnus	6·0	8·4	Dwarf	11·2	520
ε		Indus	4·7	7·0	Dwarf	11·2	469
Procyon A (α)		Canis Minor	0·3	2·6	Sub-giant	11·4	125
	B	Canis Minor	10·8	13·1	White dwarf	11·4	125

The letters A and B refer to the brighter and fainter members of a binary star system (see page 52).

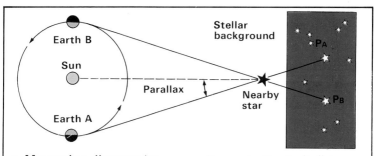

▲ **Measuring distance** by parallax. In the six months between the Earth moving from position A to position B, the star appears to shift from P_A to P_B. Knowing the distance from the Earth to the Sun (the Astronomical Unit) and the parallax angle, the star's distance may be found.

Parallax can be used for distances up to 100 light-years or so. More remote measurements are based on comparing the star's apparent magnitude with its true luminosity or absolute magnitude.

▼ **This diagram** shows the relative sizes of the Sun and its nearest companion star, as well as three planets and the Moon. To represent the relative distances, however they would have to be separated by the amounts shown, and the whole Earth could only accommodate a few of the Sun's neighbours.

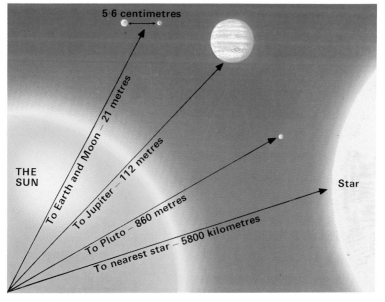

How the Stars Shine

Stars shine because of nuclear reactions deep in their interiors. They are made up mostly of hydrogen. The tremendous pressure in their centres raises the temperature to millions of degrees and in the heat the hydrogen atoms break down, recombining as helium. This releases enough energy to keep the process going as long as the hydrogen lasts.

But some stars are much hotter than others, and the colour of a star depends upon its temperature. White stars may have a surface temperature of 50,000°C; cool red stars may be below 3000°C. The Sun, at 6000°C, is yellowish.

A colour-magnitude (or Hertzsprung-Russell) diagram, shown opposite, plots surface temperature against absolute magnitude, and star families emerge. The most important is the *main sequence*, where most stars are found; they are known as *dwarfs*, to distinguish them from the inflated *giants* and *supergiants*. The *white dwarfs,* on the other hand, are small and dense, and are so dim that very few have been discovered, although they almost certainly outnumber the giants.

The colour-magnitude diagram is rather like a photograph of a crowd of people – it shows the differences between individuals. It took astronomers a long time to realize how these various types of star are related.

THE STELLAR SPECTRUM

If starlight is passed through a spectroscope — a device to broaden it into a coloured band — a number of dark lines are usually seen. The coloured band is produced by the star's shining surface, while the dark lines indicate narrow strips of colour that have been absorbed by elements in its thin surrounding atmosphere.

By matching the lines with those obtained in a laboratory, astronomers can discover what elements are present in the star's atmosphere. However, the element helium was detected in the Sun before it was discovered on the Earth.

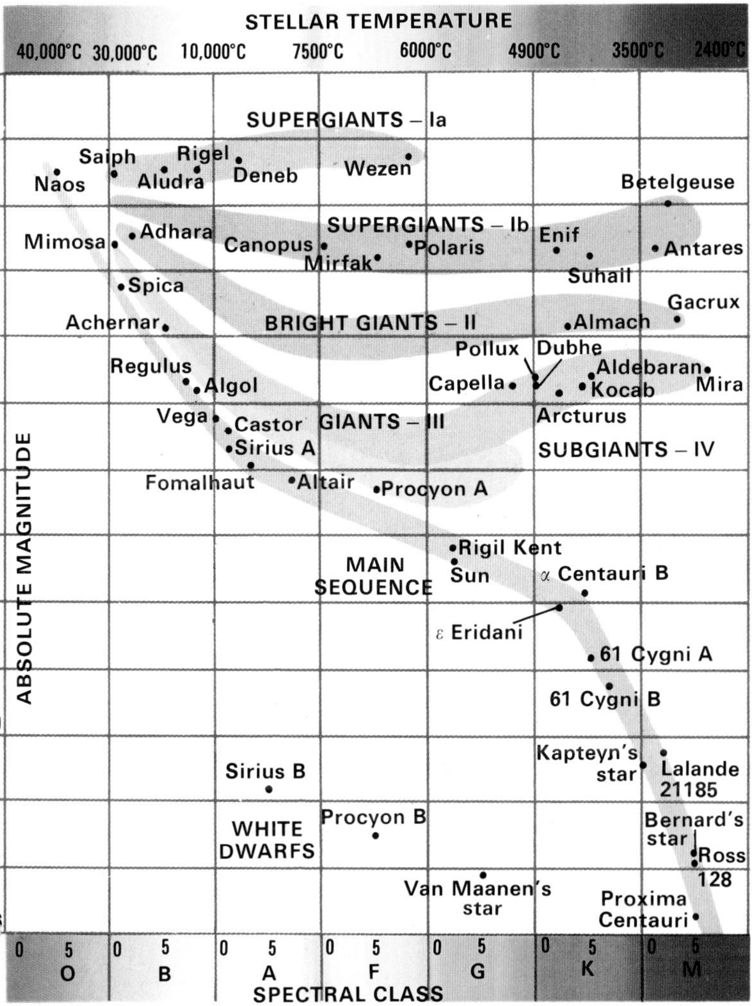

STELLAR TEMPERATURE

40,000°C	30,000°C	10,000°C	7500°C	6000°C	4900°C	3500°C	2400°C

SUPERGIANTS — Ia

Saiph Rigel
Naos Aludra Deneb Wezen Betelgeuse

Mimosa Adhara Canopus SUPERGIANTS — Ib Enif
Mirfak Polaris Antares
Spica Suhail
Achernar BRIGHT GIANTS — II Almach Gacrux

Regulus Pollux Dubhe
Algol Capella Aldebaran
Kocab Mira
Vega Castor GIANTS — III Arcturus
Sirius A SUBGIANTS — IV
Fomalhaut Altair Procyon A

ABSOLUTE MAGNITUDE

Rigil Kent
MAIN Sun α Centauri B
SEQUENCE
ε Eridani
61 Cygni A
61 Cygni B

Kapteyn's Lalande
Sirius B star 21185
WHITE Procyon B Bernard's
DWARFS star Ross
Van Maanen's 128
star Proxima
Centauri

0	5	0	5	0	5	0	5	0	5	0	5	0	5
O		B		A		F		G		K		M	

SPECTRAL CLASS

HERTZSPRUNG-RUSSELL DIAGRAM

Bright stars are often given a special name of their own, such as Sirius. Others are known by a Greek letter, or a number, followed by the Latin genitive form of constellation name. In the diagram above, for example, we find ε Eridani (star ε in the constellation Eridanus), and 61 Cygni (star 61 in the constellation Cygnus). The Greek alphabet is given on page 52. A few very unusual stars are named after the astronomer who examined them, such as Van Maanen's star, and Barnard's star.

49

The Life and Death of a Star

Stars form from the clouds of dust and gas called nebulae that make up a good proportion of the material in normal galaxies. When it becomes sufficiently dense, the nebula goes 'critical' and starts to condense into numerous clouds that are dark to begin with but heat up and begin to shine as stars. The temperature and brightness of the star depends upon the mass of the cloud.

Most stars, like the Sun, start life on the main sequence. But, as they burn their hydrogen, their cores become hotter and they emit shells of relatively cool gas: they have evolved into red giants like Aldebaran and Betelgeuse. Eventually the shell disappears and only the intensely hot white core remains: the star is now a white dwarf, like Sirius B. A brilliant star like Deneb evolves in a few million years, whereas the much dimmer Sun will remain on the main sequence for thousands of millions of years.

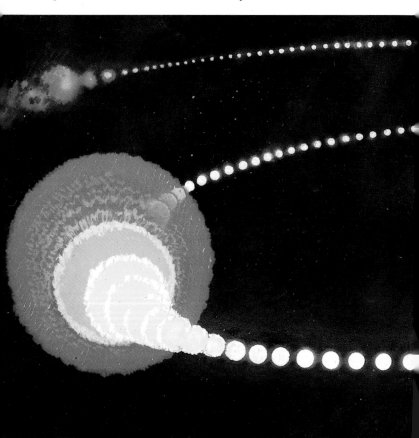

► **The Sun** is an ordinary main-sequence star. Hydrogen atoms are turned into helium atoms in its searing core, releasing enormous amounts of energy.

▼ **A normal star** condenses from a dark cloud. It shines steadily for a long time, then expands into a red giant, and dies away as a tiny white dwarf. A very massive star may end its life in a supernova explosion.

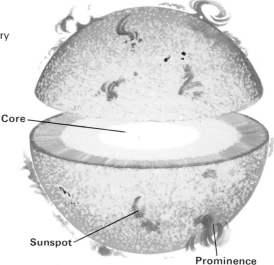

Core

Sunspot

Prominence

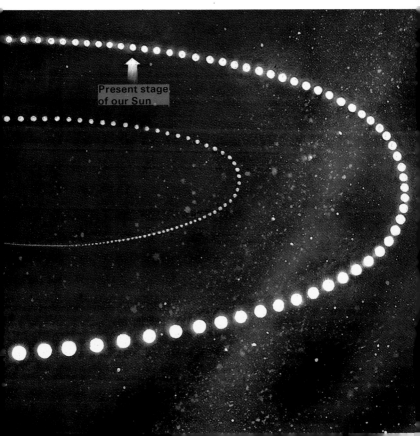

Present stage of our Sun

φ **Tauri**   η **Cassiopeiae**   β **Cygni**

Double Stars

If you look at star Zeta (ζ) in the constellation Ursa Major (Star Map 1, page 70), you will notice a much fainter star, Alcor, close to it. This is an example of a naked-eye double star. Turn a small astronomical telescope on to the pair, and Zeta itself (usually known as Mizar) is seen to be a close double.

The Mizar-Alcor pair is known as an *optical* double; the stars are far apart, and just happen to be in almost the same direction. But Mizar is a true *binary*, and the stars are revolving around each other, although they may take about 14,000 years to complete a revolution. The brighter star in a binary is known as the primary; the fainter as the secondary.

The distance between the components of a double star is measured in seconds of arc ("). There are 60" in one minute of arc ('), and 60' equal one degree. An ordinary human hair, viewed from a distance of 12 metres, is about 1" thick. Good binoculars can separate normal double stars whose components are only 30" apart. The well-known double Beta Cygni (page 86) is an excellent test object for northern observers. If one star is much brighter than the other, however, its companion will be difficult to see because it may be hidden by the glare.

LETTERING THE STARS

The brighter stars in each constellation are given a Greek letter in approximate order of brightness, beginning with α. Since a number of these bright stars are important doubles, the Greek alphabet is given here:

α alpha	η eta	ν nu	τ tau
β beta	θ theta	ξ zi	υ upsilon
γ gamma	ι iota	ο omicron	φ phi
δ delta	κ kappa	π pi	χ chi
ε epsilon	λ lambda	ρ rho	ψ psi
ζ zeta	μ mu	σ sigma	ω omega

Eclipsing Binaries

Some binaries are so close that no telescope can separate them. They can only be detected by double lines in the spectrum, or by one star eclipsing the other.

Beta Persei (Algol) is an example of an *eclipsing binary,* consisting of a bright star and a dim star. In the diagram on the right, light from both stars reaches the Earth (1). After about 18 hours, (2), the dim star partly blocks off the bright star, and the total magnitude drops. At (3), one and a half days later, there is a slight brightness drop as the dim star is obstructed.

Another naked-eye eclipsing binary, Beta Lyrae, consists of two equal stars almost touching. The light change is continuous. These two binary systems represent the *dark-eclipsing* and *bright-eclipsing* families of variable stars and hundreds of members are known.

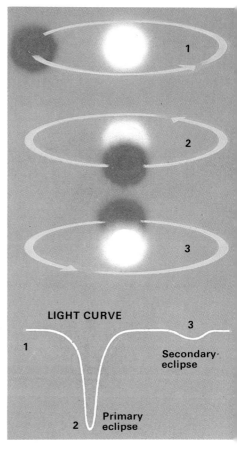

LIGHT CURVE

1

3

Secondary eclipse

Primary eclipse

2

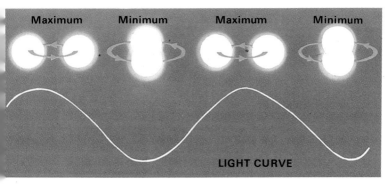

Maximum   Minimum   Maximum   Minimum

LIGHT CURVE

Variable Stars

Eclipsing binaries are known as *extrinsic variables* since the stars themselves do not change in brightness. *Intrinsic variables* actually change in luminosity usually due to huge pulsations.

One well-known group is the *Cepheid* family. These are yellow giant stars that brighten and fade by up to two magnitudes in a very regular manner. Cepheids are important, since their mean absolute magnitude is related to their period. By timing a Cepheid, its absolute magnitude can be determined. By comparing its absolute magnitude with its apparent magnitude, its distance can then be discovered.

Long-period variables (LPVs) form a very large group. They are red-giant stars with much longer periods than Cepheids. These range from about 200 to 500 days and vary by up to 10 magnitudes in brightness. They do not repeat themselves exactly from cycle to cycle. A famous long-period variable, Mira in Cetus, shows dramatic changes in its brightness; it has appeared almost as bright as the Pole Star, while at minimum it is only just visible with ordinary binoculars. LPVs usually rise to maximum brightness more rapidly than they fade afterwards.

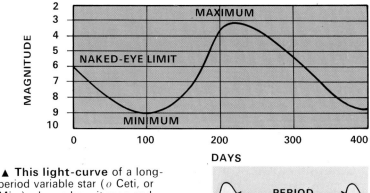

▲ **This light-curve** of a long-period variable star (*o* Ceti, or Mira), shows how it rose and fell in brightness.

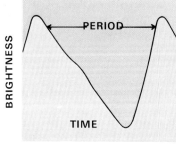

▶ **Cepheid variables** repeat their brightness variations exactly from cycle to cycle, although some vary in brightness more than others. More luminous stars have longer periods.

Observing Variable Stars

Variable-star observation is very popular with amateur astronomers. It consists of estimating the star's magnitude by comparing it with nearby *comparison stars* whose magnitude is known.

There are several ways of making an estimate. One way is to locate two comparisons, one brighter than the variable and the other fainter, and to decide how the variable 'fits' between them.

It may be exactly midway in brightness, or closer to one or the other. With skill and experience, estimates can be accurate to 0·2 magnitude or even less. This is called the *fractional* method.

You can also try the *step* method, learning to recognize steps of 0·1 magnitude, and estimating by how many steps the variable differs from a comparison star. This work is important, since professionals cannot observe all the variable stars in the sky, and some are so unpredictable that they need to be observed every night.

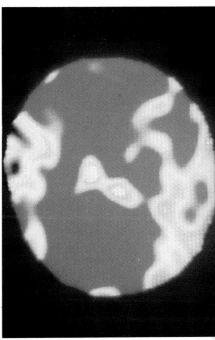

▲ **Betelgeuse,** like many red giant stars, varies slightly in brightness over the years. The image here is computer-enhanced.

▶ **This chart** shows the brighter stars in Cepheus (page 84), with comparison stars for the famous variable δ (Delta).

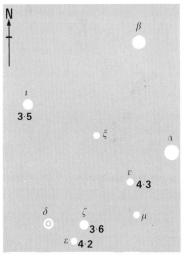

Novae and Supernovae

Some variable stars are unpredictable and violent. Every few years, a new naked-eye star suddenly shines out, literally overnight. This is a *nova,* a member of a close binary system which blasts off a huge shell of material. A nova can rise in brightness by over 10,000 times in a couple of days.

Some amateurs hunt for novae, scanning the Milky Way with binoculars. The most successful of this group, G. E. D. Alcock of Peterborough, England, has found four since 1967. To discover a nova you must know the sky really well, so that a single extra star is readily spotted.

Supernovae are rarer still. This is the complete self-destruction of a massive star, and rivals an entire galaxy in brightness. The last observed in our Galaxy was in 1604; the one of 1572 could be seen in full daylight, while the 1054 supernova has left the wreckage known as the Crab Nebula. All supernovae are shattered to pieces in their explosions, collapsing into neutron stars, smaller and denser than white dwarfs (see page 162).

▲ **The Crab Nebula** is the remains of the 1054 supernova, equivalent in violence to a million million million million hydrogen bombs.

▶ **Nova Persei** rose to naked-eye brightness in 1901. Now it is faint, with dim nebulosity.

VARIABLE STAR TYPES

Star	Magnitude range	Period	Map No.
Eclipsing binaries			
ε Aurigae	3·0 – 4·0	27·1 years	3
ζ Aurigae	3·8 – 4·3	972 days	3
β Lyrae	3·3 – 4·2	12·9 days	7
β Persei (Algol)	2·2 – 3·2	69 hours	3
Cepheid variables			
η Aquilae	4·1 – 5·4	7·2 days	7
δ Cephei	3·5 – 4·3	5·4 days	1
β Doradus	3·8 – 4·8	9·8 days	8
Long-period variables			
o Ceti (Mira)	3 – 10	330 days	2
χ Cygni	4 – 14	406 days	7
Irregular variables			
ρ Cassiopeiae	4 – 6	—	1
μ Cephei	4 – 5	—	1
α Herculis (Rasalgethi)	3 – 4	—	6
α Orionis (Betelgeuse)	0·4 – 1·3	5 years?	3
α Scorpii (Antares)	0·9 – 1·8	5 years?	6
Other variables			
γ Cassiopeiae	1·7 – 2·4	(Usually faint)	1
T Coronae Borealis	2 – 10	(Usually faint)	6
R Coronae Borealis	6 – 14	(Usually bright)	6

Nebulae and Star Clusters

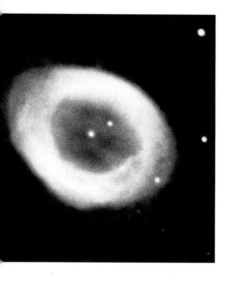

The Ring Nebula (left) was expelled from a star. The Horsehead (above) and Orion Nebula (right) indicate where young stars are forming.

The word *nebula* means a cloud, and there are many cloudy-looking objects in the sky. But any comparison with terrestrial clouds is misleading. A cubic metre of rain cloud weighs perhaps 100 grammes, but a volume of nebula the size of the Earth would weigh only a few kilogrammes! Nebulae are noticeable only because they are so huge.

A nebula, like almost everything in the universe, consists mostly of hydrogen, although compounds of hydrogen with nitrogen, carbon, oxygen and other elements may be found. There are four main kinds of nebula:

Planetary nebulae are shells of gas expelled from very hot stars. They are called planetary nebulae because some of them appear disc-like in small telescopes. They are usually relatively small – less than a light-year across. Most are very faint. An example is M57 in Lyra (Star Map 7).

Reflection nebulae shine by reflected starlight, and are usually faint. An example is the Pleiades in Taurus (Star Map 3).

Emission nebulae are huge irregular clouds tens of light-years across. Within them stars are forming. They shine because their atoms react with radiation from nearby hot stars. An example is M8 in Sagittarius (Star Map 7).

Dark nebulae can be detected only by what they obscure. The Milky Way outline is irregular because of dark nebulae in front of the stars. An example is M42 in Orion (Star Map 3).

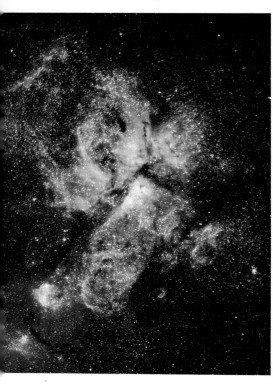

◄ **The Keyhole Nebula** is at this moment giving birth to a cluster of stars.

► **The Pleiades** are a young cluster, probably less than 100 million years old. The six or seven brightest stars are detectable with the naked eye: they are white giant stars thousands of times as luminous as the Sun. They have not yet had time to evolve. As yet, there are neither red giants nor white dwarfs in the Pleiades.

Star Clusters

Stars form in clusters rather than as individuals, condensing out of separate clouds within a nebula many light-years across. The Sun, presumably, was once a cluster member. But every star has a separate motion of its own, and unless the cluster is very compact so that the gravitational pull of its members holds the group together, the stars float apart over many millions of years.

These *open* clusters are being born all the time. For example, young stars are being formed in the Orion Nebula, M42. Clusters are very interesting to astronomers, since the stars in them form a mixed group of types, all of the same age. About 300 open clusters have been discovered in the Galaxy.

A cluster's likely age can be determined by examining its population. If it contains many very hot stars, it must be young – since such stars burn out quickly, or evolve into red giants. If it contains red giants and white dwarfs, it must be old.

Open clusters occur throughout the arms of the Galaxy, but *globular* clusters are completely different. They are about as old as the Galaxy itself, and contain red-giant stars. They measure up to 100 light-years across, and their population is reckoned in hundreds of thousands of stars. There are over 100 globular clusters in the Galaxy and many more in some other galaxies. They form a 'halo' around our star system.

Observing Star Clusters

The brightest open and globular clusters can be seen with the naked eye, provided the sky is really dark. Binoculars will give a good view of some of the larger open clusters. With apertures of from 60 millimetres to 150 millimetres, many are superb. Even experienced amateurs never tire of these objects.

You should remember that photographs taken with large telescopes show far more detail in clusters and nebulae than may be visible to the naked eye, even when using a large instrument. On the other hand, no photograph can capture the telescopic brilliance of stars sparkling in the eyepiece.

▶ **The Great Cluster** in Hercules (top) is one of the nearer globular clusters, and may contain half a million stars.

▶ **Omega Centauri**, the finest globular cluster in the sky, lies well south of the celestial equator. It is easily visible with the naked eye.

◀ **The Coal Sack** nebula obscures stars in the Milky Way.

63

The Constellations

The constellations are imaginary groupings of the stars. Invented years ago by people to help map the sky, they are still the easiest way to learn the stars. This chapter describes the most interesting of the 88 constellations, all of which are shown on the maps. Their brighter or more important stars are identified by name or letter. The standard three-letter abbreviation of each constellation is also given with each entry.

Nebulae and clusters, as well as galaxies, are given an 'M' number from Messier's 1781 catalogue (for example the Orion Nebula, M42), or a number alone from the New General Catalogue (NGC) of 1888 (for example the Double Cluster in Perseus, 869 and 884).

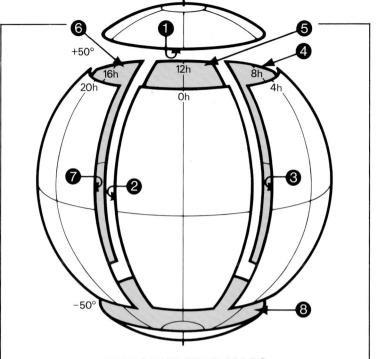

USING THE STAR MAPS

The maps on pages 70-79 represent the whole celestial sphere divided into six segments and the northern and southern circumpolar regions as shown above. The areas covered by each map are as follows:

1 Dec +50° to N pole
2 RA 22h – 2h ⎫
3 RA 2h – 6h ⎪
4 RA 6h – 10h ⎬ Dec +50°
5 RA 10h – 14h ⎪ to –50°
6 RA 14h – 18h ⎪
7 RA 18h – 22h ⎭
8 Dec –50° to S pole

The maps show all the stars in the sky down to about magnitude 4·5 as well as the positions of clusters, nebulae, and so on, usually known as 'deep-sky objects'.

These deep-sky objects are usually much fainter than the stars shown on the maps. Some special charts, to help locate the more difficult objects, have been added to the constellation notes.

There are also special charts for variable stars, which give the magnitudes of suitable comparison stars.

The light blue strip shows the path of the Milky Way and some of the more obvious irregularities in its course, many of which are caused by dark nebulae.

Finding a Star

Star maps can at first be very confusing. It is difficult to compare the stars shown flat on paper to the arc of stars in the sky above. A good way of avoiding disappointment is to learn to look for some of the brighter stars. Once you are familiar with these, the other patterns will be easier to find.

The constellations Ursa Major and Orion are so well known that they can be used to identify other stars and groups. Ursa Major is always in the northern part of the sky, while Orion lies on the celestial equator. It is best seen between September and April. The region of Ursa Major is shown on Star Maps 1 and 5, while Orion appears on Map 3.

The diagrams on this page show how to use different star alignments in these two constellations. If you find it difficult to sort out the sky in this way, use the systematic method described below.

Where to Look

All celestial objects are highest when crossing the *meridian* (the north-south line that passes overhead). Unless the object is near the celestial pole, it will be due south to a northern observer, and due north to a southern one.

URSA MAJOR

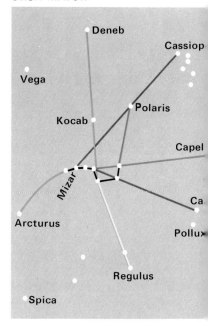

ORION

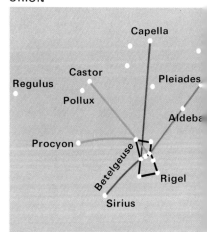

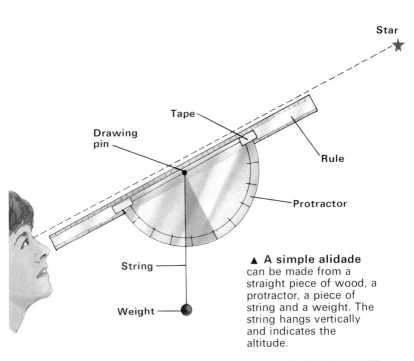

Star

Tape

Drawing pin

Rule

Protractor

String

Weight

▲ **A simple alidade** can be made from a straight piece of wood, a protractor, a piece of string and a weight. The string hangs vertically and indicates the altitude.

Take the Dec of the brightest star in the constellation from the notes that follow the star maps. Also find out the co-latitude of your site (the co-latitude is equal to 90 degrees (°) minus your latitude).

Add the star's Dec to this amount, and the result is the star's altitude above the horizon as it crosses the meridian. This is the angle to which you must set your alidade.

Don't forget that if the star has a southern Dec, you subtract the number from the complement. If you live south of the Equator, call south Dec positive and north Dec negative and add and subtract accordingly.

CLOTHING AND EQUIPMENT

It may not feel cold when you step outside – but even summer evenings can quickly chill enthusiasm if you are not properly clothed.

Good dark-adaptation is also essential. The eyes take at least ten minutes to acclimatize, so make sure that you have not left anything vital in a brightly-lit room.

Work with a dim red light. A bicycle rear lamp may be rather too bright. Try painting the bulb of an ordinary torch with red poster paint.

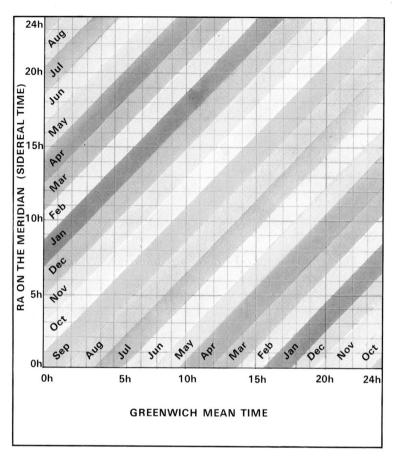

Greenwich Mean Time axis labeled 0h, 5h, 10h, 15h, 20h, 24h

RA ON THE MERIDIAN (SIDEREAL TIME) axis labeled 0h, 5h, 10h, 15h, 20h, 24h

GREENWICH MEAN TIME

Timing Observations

As well as knowing *where* to look, you must know *when* to look. As the Earth turns, objects are carried past the meridian. The Sidereal Time (ST) gives the RA that is on the meridian at any moment.

For example, Regulus (Alpha Leonis) is at RA 10h 06m, so it is on the meridian just after 10h

ST. To convert ST to GMT, use the converter above. For example, if you are observing on January 31, 10h 06m is about the same as 01h 30m GMT, so Regulus will be on the meridian at 1.30 am. (Note that while GMT is used by British observers, in other parts of the world you must use your own local time system.) A better

68

time to observe Leo would be in March; on March 31, for example, Regulus will be on the meridian at 21.30 GMT (9.30 pm).

Once you have identified a few of the brighter groups, the fainter constellations can be fitted into the gaps quite easily, using the maps in this book. Do not be content with learning just the brighter stars. Take a constellation and study it, using an atlas such as *Norton's* (see page 183) until you know the letters of all the naked-eye stars. If it is a Milky Way constellation, keep an eye out for a nova.

Very few amateurs, even experienced ones have ever bothered to learn more than about 20 bright 'signpost' stars. But if you make an effort and observe systematically, you can memorize hundreds of stars.

Finally, buy a notebook and record all you learn and see (or fail to see!). You will come to treasure it later, and some of the information may be valuable in years to come. Include the date and time with all observations.

▼ One way to set out your observation book.

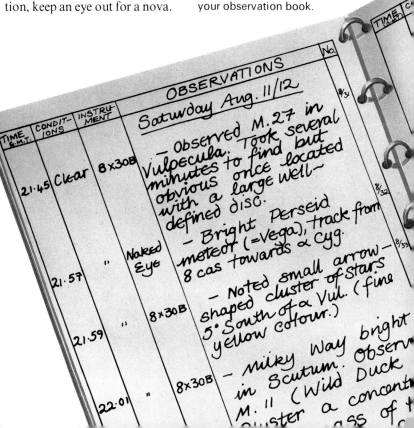

Star Map 1 –
Northern Circumpolar Stars

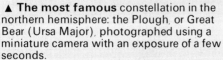

▲ **The most famous** constellation in the northern hemisphere: the Plough, or Great Bear (Ursa Major), photographed using a miniature camera with an exposure of a few seconds.

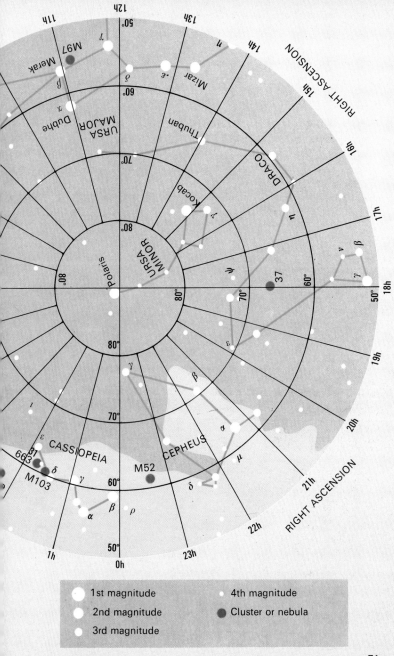

1st magnitude
2nd magnitude
3rd magnitude
4th magnitude
Cluster or nebula

71

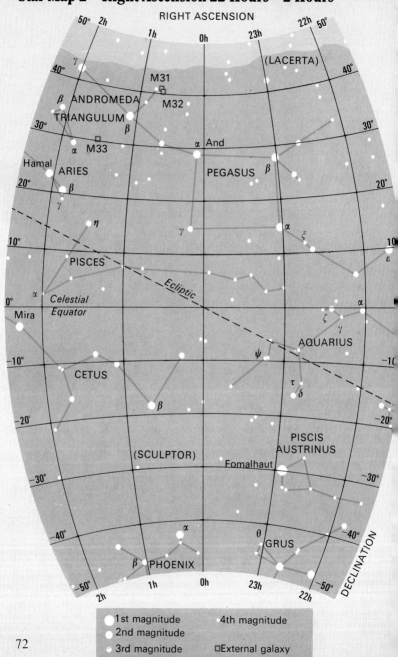

Star Map 2 – Right Ascension 22 Hours – 2 Hours

RIGHT ASCENSION

(LACERTA)

ANDROMEDA

M31

M32

TRIANGULUM

And

M33

PEGASUS

Hamal

ARIES

PISCES

Ecliptic

Celestial
Equator

Mira

AQUARIUS

CETUS

PISCIS
AUSTRINUS

(SCULPTOR)

Fomalhaut

PHOENIX

GRUS

DECLINATION

1st magnitude
2nd magnitude
3rd magnitude

4th magnitude

□External galaxy

Star Map 3 – Right Ascension 2 Hours – 6 Hours

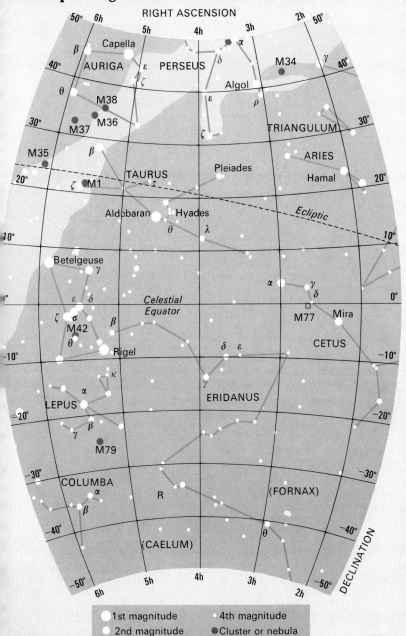

RIGHT ASCENSION

AURIGA
Capella
β
θ
ε
ζ
M38
M37 M36
β
M35
ζ M1
TAURUS
Aldebaran θ
Hyades
λ

PERSEUS
α
δ
Algol
ε
ζ
M34
γ
ρ
TRIANGULUM
ARIES
Hamal
Pleiades
τ
Ecliptic

Betelgeuse
γ
ε δ
Celestial Equator
ζ σ
M42
θ
β
Rigel
κ
α
LEPUS
γ β
M79
COLUMBA
α
β
(CAELUM)

α
γ δ
M77 Mira
CETUS
δ ε
ERIDANUS
γ

R
(FORNAX)
θ

DECLINATION

1st magnitude
2nd magnitude
3rd magnitude
4th magnitude
Cluster or nebula
External galaxy

73

Star Map 4 – Right Ascension 6 Hours – 10 Hours

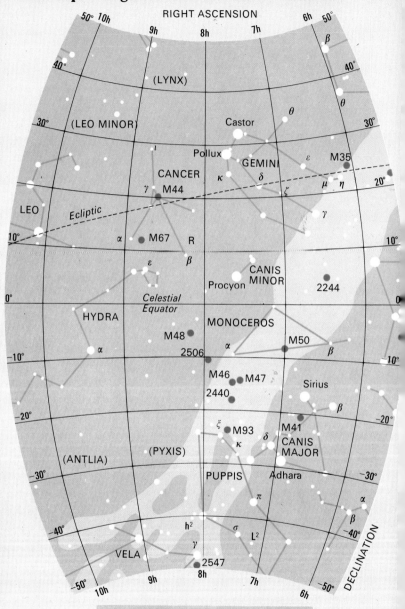

RIGHT ASCENSION

DECLINATION

○ 1st magnitude	· 4th magnitude
○ 2nd magnitude	● Cluster or nebula
○ 3rd magnitude	

74

Star Map 5 – Right Ascension 10 Hours – 14 Hours

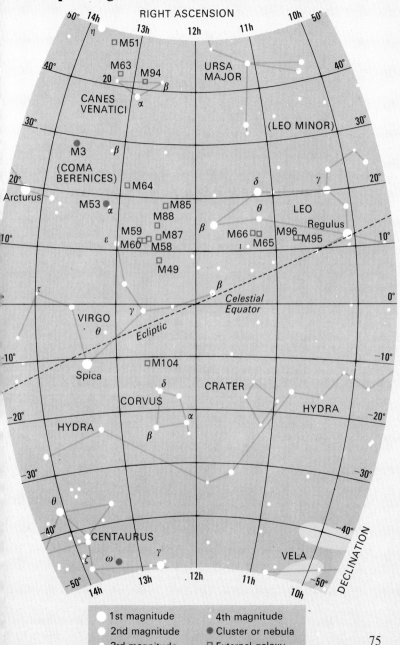

RIGHT ASCENSION

URSA MAJOR

(LEO MINOR)

CANES VENATICI

M51
M63
M94
20
β
α

M3
β
(COMA BERENICES)

Arcturus

M53
α

M64
M85
M88
M59
M87
M60
M58
M49
ε

δ
γ

θ
LEO
M66
M65
M96
M95
Regulus
ι
β

τ
γ
β
Celestial Equator

VIRGO
θ
Ecliptic

Spica
M104

CORVUS
δ
CRATER
α
β

HYDRA
HYDRA

θ

CENTAURUS
ζ
ω
γ

VELA

DECLINATION

1st magnitude
2nd magnitude
3rd magnitude
4th magnitude
Cluster or nebula
External galaxy

75

Star Map 6 – Right Ascension 14 Hours – 18 Hours

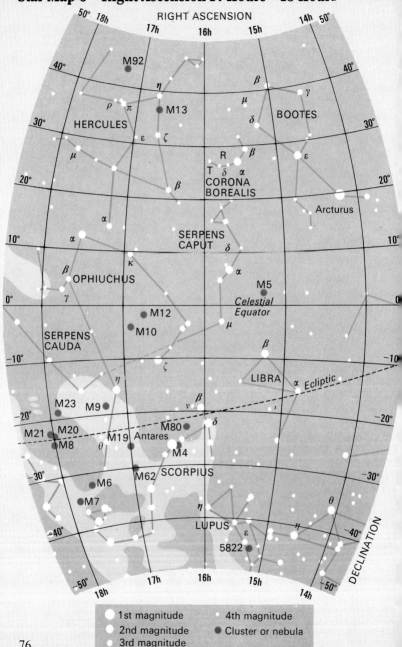

RIGHT ASCENSION

50° 18h 17h 16h 15h 14h 50°

40° M92 40°

η β

30° ρ π M13 μ BOOTES γ 30°
HERCULES δ
ε ζ

R β
μ 20° T δ α ε Arcturus 20°
β CORONA
BOREALIS

α SERPENS
10° α CAPUT δ 10°
β κ α
OPHIUCHUS M5
0° γ Celestial 0°
M12 Equator
M10 μ
-10° SERPENS -10°
CAUDA ζ
LIBRA α Ecliptic
η ν β
-20° M23 M9 M80 -20°
M21 M20 δ
M8 M19 Antares
θ M4
-30° M62 SCORPIUS -30°
M6
M7 θ
-40° η -40°
LUPUS ε
5822
-50° 18h 17h 16h 15h 14h -50°

DECLINATION

○ 1st magnitude   · 4th magnitude
○ 2nd magnitude   ● Cluster or nebula
○ 3rd magnitude

76

Star Map 7 – Right Ascension 18 Hours – 22 Hours

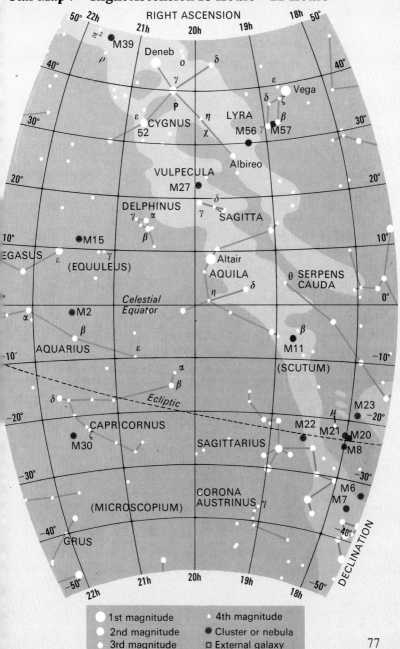

RIGHT ASCENSION

1st magnitude
2nd magnitude
3rd magnitude
4th magnitude
Cluster or nebula
External galaxy

77

Star Map 8 –
Southern Circumpolar Stars

RIGHT ASCENSION

15h

18h

(NORMA)

17h

β

ARA

−50°

−60°

18h

6752

19h

20h

α

21h

RIGHT ASCENSION

▲ **The Southern Cross**, a fabulous
sight among the southern circumpolar
stars.

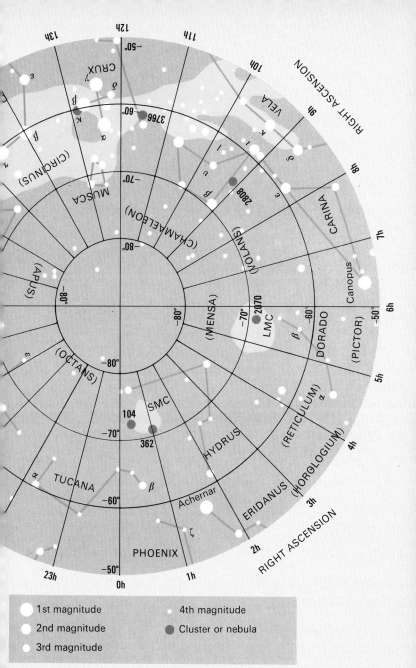

Observation Notes on the Constellations

All the stars and other objects mentioned in these notes can be found on the Star Maps and should be visible with binoculars, although the nebulae and clusters will look much more impressive if a powerful telescope is available.

Andromeda (And, Star Map 2) This large constellation contains the famous Andromeda Galaxy, M31, easily seen with the naked eye in a dark sky; binoculars

▼ **The globular cluster** M2 in Aquarius is one of the most remote objects in the Galaxy visible with binoculars.

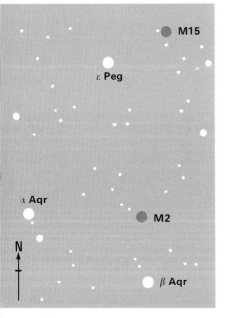

show a small hazy companion, M32, both about two million light-years away. α (Alpheratz): RA 00h 06m, Dec +29°, mag 2·0.

Apus, the Bird of Paradise
(Aps, Star Map 8)
A very faint group near the south pole. α : RA 14h 42m, Dec –79°, mag 3·8.

Aquarius, the Water Bearer
(Aqr, Star Maps 2 and 7)
This zodiacal constellation contains only two bright stars, but the little group of faint stars around ζ has a distinctive shape. Use the special chart to locate the globular cluster M2, which appears in binoculars as a very small hazy spot (diameter about 100 light-years; distance 55,000 light-years). The mag 4 star τ is a wide binocular double. α (Sadalmelik): RA 22h 03m, Dec $-\frac{1}{2}°$, mag 2·9 – one of the closest bright stars to the celestial equator.

Aquila, the Eagle (Aql, Star Map 7)
A notable Milky Way group whose brightest star, Altair, is one of the Sun's nearer neighbours, only 16 light-years away. η Aql is a bright Cepheid (see page 54); fine star fields near δ. α (Altair): RA 19h 48m, Dec $+8\frac{1}{2}°$, mag 0·8.

▲ **The star-field** around
Altair in Aquila. Most of these
stars are more luminous than
the Sun.

Ara, the Altar (Ara, Star Map 8)
A conspicuous, compact group
in the southern Milky Way.
β: RA 17h 21m, Dec $-55\frac{1}{2}°$,
mag 2·8.

Aries, the Ram (Ari, Star Map 3)
A small and rather faint group,
but noticeable because there are
few stars nearby. α (Hamal):
RA 2h 04m, Dec $+23°$, mag 2·0.

Auriga, the Charioteer (Aur, Star
Map 3)
A magnificent constellation
lying in a rather faint part of the
northern Milky Way. Its three
prominent open clusters, M36,
M37 and M38, are all fine
binocular objects. ε and ζ are
both dark-eclipsing variable
stars (see page 53): the next
minimum of ε was due to begin in
July 1982. α (Capella): RA 5h
13m, Dec $+46°$, mag 0·1.

Boötes, the Herdsman (Boo, Star
Map 6)
A prominent kite-shaped group,
with the brilliant reddish Arcturus in its 'tail'. δ has a mag 9
companion almost $2'$ to the east,
but there are few deep-sky objects. α (Arcturus): RA 14h
13m, Dec $+19\frac{1}{2}°$, mag $-0·1$.

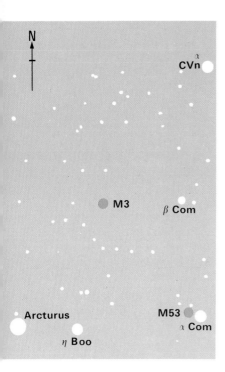

N

CVn
M3
β Com
Arcturus
M53
λ Com
η Boo

▲ **M3**, a bright globular cluster in Canes Venatici, lies on a line from α CVn towards Arcturus in Boötes.

Cancer, the Crab (Cnc, Star Map 4)
A faint zodiacal group, interesting for the bright cluster Praesepe (M44), which looks like a large hazy patch with the naked eye. M44 must be much older than the Pleiades in Taurus because it contains no luminous white stars; they are all yellowish main-sequence dwarfs, or red giants. This group is about 15 light-years across and 500 light-years away. M67 is even older,

and appears in binoculars as a misty spot. β: RA 8h 14m, Dec $+9\frac{1}{2}°$, mag 3·5.

Canis Major, the Greater Dog (CMa, Star Map 4)
A brilliant Milky Way group, containing the brightest star in the sky, Sirius. Viewed from north European latitudes, Sirius never rises very high above the southern horizon, usually twinkling violently through the unsteady atmosphere. The open cluster M41 is easy to locate with binoculars. α (Sirius): RA 6h 43m, Dec $-16\frac{1}{2}°$, mag $-1·5$.

Canis Minor, the Lesser Dog (CMi, Star Map 4)
Only its leading star is obvious; like Sirius, it is a binary with a white dwarf companion. α (Procyon): RA 7h 37m, Dec $+5\frac{1}{2}°$, mag 0·3.

Canes Venatici, the Hunting Dogs (CVn, Star Map 5)
Only the leading star is obvious, lying south of Ursa Major. Use the accompanying chart to locate M3, one of the finest globular clusters. The external galaxy M51 may also be glimpsed as a very faint hazy disc. α (Cor Caroli): RA 12h 54m, Dec $+38\frac{1}{2}°$, mag 2·8.

Capricornus, the Sea Goat (Cap, Star Map 7)
A large but faint zodiacal constellation, containing the wide

naked-eye double star α (mags 3·6 and 4·3, 6′ 16″ apart), and an easy binocular double, β, which has a mag 6 companion 3′ 25″ away. The globular cluster M30 is a difficult binocular object from high northern latitudes. α (Dabih): RA 20h 18m, Dec −15°, mag 3·0.

Carina, the Keel (Car, Star Map 8)

A large and brilliant southern constellation containing several bright clusters and groups, including the globular cluster NGC 2808, visible with binoculars. Magnificent sweeping in

▼ **This 1-minute** exposure with a fixed camera shows Cassiopeia, part of Perseus, and the Double Cluster between them.

the Milky Way near Vela. α (Canopus): RA 6h 23m, Dec −52½°, mag −0·7.

Cassiopeia (Cas, Star Map 1)

A small but unmistakable constellation, its five brightest stars forming a conspicuous M or W near the north celestial pole. γ is an unusual variable, which suddenly brightened to mag 1·7 from its usual 2·4 in 1938. ρ, usually mag 5, sometimes fades to 6. There are rich Milky Way fields; note the clusters M52 and NGC 663 particularly. α (Shedir): RA 0h 38m, Dec +56½°, mag 2·2.

Centaurus, the Centaur (Cen, Star Maps 5, 6 and 8)

An extensive constellation. α is a

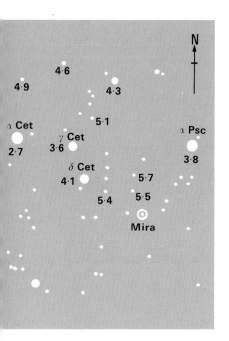

◄ **If the long-period** variable
Mira (*o* Ceti) is above magnitude
6, it can be located with this
chart and followed as it
brightens or fades.

(see page 55 and use the chart
given there to observe its
changes). It has a mag 5 companion 41″ away. α (Alder-
amin): RA 21h 17m, Dec +62½°,
mag 2·4.

Cetus, the Whale
(Cet, Star Maps 2 and 3)
An extensive faint constellation,
although the little triangle of α,
γ and δ is fairly obvious. Mira,
or *o*, is a long-period variable
mentioned on page 54, ranging
from about mag 10 at minimum
to as bright as mag 2 (but usually
about mag 4) at maximum, in a
period of about 11 months. The
next maxima should occur ap-
proximately in June 1983 and
May 1984, when it will be an
early-morning object. α (Men-
kar): RA 3h 00m, Dec +4½°,
mag 2·7.

Coma Berenices, Berenices' Hair
(Com, Star Map 5)
Although containing no star
brighter than mag 4, Coma can
be distinguished as a faint scat-
tering of stars in an otherwise
dull patch of sky. It contains one
of the largest clusters of galaxies
known – several thousand of
them – but only a few are
prominent. The chart shows the
positions of two that are just

magnificent binary star although
it cannot be resolved without a
proper telescope. Known as
Rigel Kent, it is the closest
naked-eye star to the Sun. Centaurus contains the closest and
brightest globular cluster in the
sky, labelled ω. Binoculars show
it as an immense hazy patch
about two-thirds the Moon's
diameter. α (Rigel Kent): RA
14h 36m, Dec –60½°, mag 0·1.

Cepheus (Cep, Star Map 1)
Not easy to identify, but contains interesting objects. μ was
called the Garnet Star by
Herschel because of its port-
wine colour; it is slightly variable. δ is the prototype Cepheid

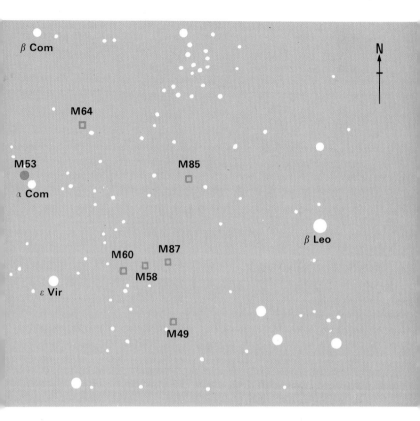

N

β Com

M64

M53

α Com

M85

β Leo

M60   M87

M58

ε Vir

M49

visible with binoculars (M64 and M85); M53 is brighter than either, but is a globular cluster. Both galaxies are about 40 million light years away. α RA 13h 08m, Dec +18°, mag 4·2.

Corona Borealis, the Northern Crown (CrB, Star Map 6)
Although small, this group is easily identified from its semicircular outline. Note variable R, normally mag 6, which at unpredictable intervals sinks in a few days to as faint as mag 14;

▲ **These are** some of the brighter galaxies in the Coma-Virgo region, together with the globular cluster M53, the easiest Messier object shown here.

see the chart on page 86. T rose to brief naked-eye visibility in 1866 and 1946. α (Alphecca): RA 15h 33m, Dec +27°, mag 2·3.

Cygnus, the Swan (Cyg, Star Map 7)
One of the finest northern constellations, the Milky Way about γ being particularly rich. It

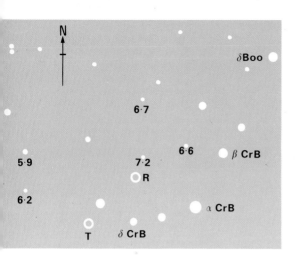

◀ **A chart** for the unusual variables R and T Coronae Borealis.

contains the bright cluster M39, but the stars are too scattered for the best effect. *o* is one of the finest binocular pairs, the mag 4 yellow star making the mag 5 companion look blue. *β* (mags 3 and 5·5, separation 34″) may just be divided. Note also the long-period variable *χ*, which at maximum reaches mag 4·5, and can be identified by its reddish colour. Its period is about 406 days. *α* (Deneb): RA 20h 40m, Dec +45°, mag 1·3.

Dorado, the Dolphin
(Dor, Star Map 8)
A far southern constellation, interesting mainly because it contains one of the Galaxy's satellites, the Large Magellanic Cloud, about 160,000 light-years away and looking like a large hazy patch. This contains many superb objects including NGC 2070, a naked-eye emission nebula. *β* is a bright Cepheid (see page 54). *α*: RA 4h 33m, Dec –55°, mag 3·3.

Draco, the Dragon
(Dra, Star Map 1)
A winding constellation extending around a large arc of the northern sky. The two stars *β* and *γ* are the easiest to identify. Nearby is *ν*, a very attractive pair of mag 5 stars about 1′ apart. *α* (Thuban): RA 14h 03m, Dec +64½°, mag 3·6.

Equuleus, the Little Horse
(Equ, Star Map 7)
This tiny constellation contains only three obvious stars. One of them, *γ*, is an attractive wide double, mags 4·5 and 6. *α*: RA 21h 13m, Dec +5°, mag 3·9.

Eridanus (Eri, Star Maps 3 and 8)
A winding constellation extending from the far southern sky to

► **In Gemini,** A is the Sun's position at northern midsummer, B is where Uranus was found in 1781.

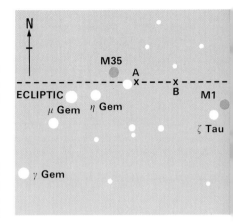

equatorial latitudes. Note the reddish tint of γ, a red giant star. α (Achernar): RA 1h 36m, Dec $-57\frac{1}{2}°$, mag 0·5.

Gemini, the Twins
(Gem, Star Map 4)
One of the most interesting of all constellations. It lies on the most northern part of the ecliptic, and the Sun moves into it from adjacent Taurus on the first day of northern summer, June 21. At this time it is only a degree or so away from the bright open cluster M35. This is

also the point where Herschel discovered the planet Uranus in 1781. μ and η both have lovely golden tints, while ε and ζ have faint binocular companions, the latter being the more impressive. β (Pollux): RA 7h 42m, Dec $+28°$, mag 1·1.

Hercules (Her, Star Map 6)
Not obvious until the quadrilateral or 'keystone' of π, η, ζ and ε is identified, but a fine constellation for the double-star enthusiast armed with an adequate telescope. M13 is its most famous object: one of the brightest globular clusters in the sky, and just visible with the naked eye. It is about 25,000 light-years away and may contain half a million stars. M92, another globular, is harder to find because there are no nearby

◄ **Dense star-field** near γ Cygni.

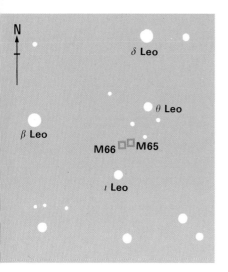

planets can occult it. Leo lies near the huge groups of galaxies in Coma and Virgo, and at least two objects, M65 and M66, can be made out with binoculars if the sky is dark; use the accompanying chart to locate them. α (Regulus): RA 10h 06m, Dec +12°, Mag 1·4.

Lepus, the Hare
(Lep, Star Map 3)
A small but conspicuous group south of Orion. γ is a very fine binocular double (mags 4 and 6·5, separation 1½′), while the globular cluster M79 can be seen

▲ **The pair** of spiral galaxies M65 and M66, in Leo.

▶ **The constellation** Lyra with its bright star Vega.

bright stars as guides; it is half as far away again as M13. α (Rasalgethi): RA 17h 12m, Dec +14½°, mag 3·5 (slightly variable).

Hydra, the Water Serpent
(Hya, Star Maps 4 and 5)
A faint, rambling group; only Alphard and the little collection of stars south of Cancer, marking its head, are obvious. α (Alphard): RA 9h 25m, Dec −8½°, mag 2·0.

Leo, the Lion (Leo, Star Map 5)
A zodiacal constellation, α lying within a degree of the ecliptic, so that both the Moon and the

as a hazy spot. α (Arneb): RA 5h 31m, Dec –18°, mag 2·6.

Libra, the Scales
(Lib, Star Map 6)
A large zodiacal constellation, but only α and β are obvious. α has a wide mag 6 companion. α (Zubenelgenubi): RA 14h 48m, Dec –16°, mag 2·8.

Lupus, the Wolf
(Lup, Star Map 6)
A small but prominent group in a rich region of the southern Milky Way. η is a mag 4 star with a mag 9 companion 2′ away, and NGC 5822 is a bright open cluster. α: RA 14h 39m, Dec –47°, mag 2·3.

Lyra, the Lyre (Lyr, Star Map 7)
A small but most interesting group. ε can be separated by acute naked eyesight; ζ (mag 4) has a mag 6 companion 44″ away. δ is another naked-eye pair. β is an eclipsing binary (see page 53): it is almost as bright as its neighbour when at its brightest, but a magnitude dimmer when passing through an eclipse. M57, the famous Ring Nebula, is barely detectable with binoculars; M56 is a faint globular. α (Vega): RA 18h 35m, Dec + 38½°, mag 0·0.

▶ **A chart** for the eclipsing star β Lyrae, also showing the position of M57, the Ring Nebula.

Monoceros, the Unicorn
(Mon, Star Map 4)
Most of the stars in this group are faint, but the Milky Way is magnificent. The cluster NGC 2244 can be seen with the naked eye. M50 and NGC 2506 are other fine clusters. M50 is so compact that it appears as a hazy patch when viewed with binoculars. α: RA 7h 39m, Dec –9½°, mag 3·9.

Ophiuchus, the Serpent Bearer
(Oph, Star Map 6)
Although not one of the twelve zodiacal constellations, the ecliptic passes through it. Superb Milky Way fields to the south; M10 and M12 are bright globular clusters, while M9, M19 and M62 are fainter globulars. The whole region is well worth sweeping on a clear night. An obvious binocular group of 8th magnitude stars about a degree away from β was not, strangely

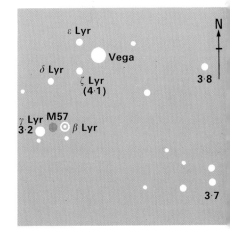

◀ **This photograph** shows the three 'belt' stars of Orion, with the Great Nebula, M42, to the south.

binocular double, with a mag 6·5 companion to the north 53″ away. The famous Great Nebula (M42) shows many irregularities due to dark nebulae. β (Rigel): RA 5h 12m, Dec $-8\frac{1}{2}°$, mag 0·1.

Pavo, the Peacock
(Pav, Star Map 8)
This group contains two interesting stars: κ, a Cepheid (mag 3·9 – 4·9, period 9·1 days), and λ, an irregular variable, mag range 3·5–4·5. NCG 6752 is a huge globular cluster, over half the Moon's diameter across. α: RA 20h 22m; Dec $-57°$, mag 1·9

Pegasus (Peg, Star Maps 2 and 7)
The Great Square is easy to recognize once located, but it is not always easy to find at first because it appears larger in the sky than on a map. ε is an easy binocular double (mags 2·5 and 8·5, separation almost 2′), and this acts as a guide to M15, a globular cluster easily seen in binoculars. α (Markab): RA 23h 02m, Dec $+15°$, mag 2·5.

Perseus (Per, Star Maps 1 and 3)
A magnificent constellation in the northern Milky Way. Its most famous object is the eclipsing binary β (Algol), described on page 53. NGC 869 and 884 form the Double Cluster, visible

enough, included in the early catalogues of deep-sky objects. α (Rasalhague): RA 17h 33m, Dec $+12\frac{1}{2}°$, mag 2·1.

Orion (Ori, Star Map 3)
This brilliant constellation contains no less than seven 1st magnitude stars – more than any other group. All, Betelgeuse apart, form a true association in space, being hot young stars. Rigel (β) is one of the most luminous known stars in the Galaxy, 50,000 times as bright as the Sun. δ is a difficult

with the naked eye and a fine binocular object. M34, a more scattered cluster, can also be seen with the naked eye. α (Mirfak): RA 3h 21m, Dec +50°, mag 1·8.

Pisces, the Fishes
(Psc, Star Map 2)

A faint, straggling zodiacal group, containing the vernal equinox – the point where the Sun crosses the celestial equator at the beginning of northern spring. α (Kaitain): RA 1h 59m, Dec +2½°, mag 3·8.

▼ **The Double Cluster** in Perseus lies in a rich region of the Milky Way.

Piscis Austrinus, the Southern Fish (PsA, Star Map 2)

A small group marked by its leader Fomalhaut, which shines in an empty part of the sky. α (Fomalhaut): RA 22h 55m, Dec –30°, mag 1·2.

Puppis, the Poop
(Pup, Star Map 4)

This small bright group contains many open clusters and fine Milky Way fields. Note particularly M47, a magnificent loose group almost equal to the Moon's diameter, and M93. M46 is rather faint for binocular observation. ζ: RA 8h 02m, Dec –40°, mag 2·3.

Sagitta, the Arrow
(Sge, Star Map 7)

A small but distinctive group lying in the Milky Way. Look just south of the midway point between γ and δ for M71 (see chart on page 96). Appearing as a dim hazy spot, it is either a very compressed open cluster or a very unusual globular. γ: RA 19h 57m, Dec $+19°$, mag 3·5.

Sagittarius, the Archer
(Sgr, Star Map 7)

A superb zodiacal constellation lying in the densest part of the Milky Way – the region towards the centre of the Galaxy. Clusters and nebulae are strewn so thickly that no guide to interesting objects is really necessary, and the ones shown on the map are only a few of the more obvious. M8, the Lagoon Nebula, is a bright cluster enveloped in haze, while M20 is the Trifid Nebula, a prominent irregular patch next to the open cluster M21. M23 is another bright cluster, while M22 is a superb globular. The constellation contains 15 Messier objects altogether, but most are poorly seen from high northern latitudes. ε (Kaus Australis): RA 18h 21m, Dec $-34\frac{1}{2}°$, mag 1·8.

◄ **The Trifid Nebula** is about 2300 light-years away from the Sun. The lanes are due to huge dark interstellar clouds between us and the nebula.

Scorpius, the Scorpion
(Sco, Star Map 6)

After Orion, perhaps the most brilliant constellation in the sky, with red Antares in its head and a curved 'sting' behind. Like its neighbour Sagittarius, Scorpius contains the densest part of the Milky Way and clusters and nebulae abound, but there are also some interesting stars. Antares is a red supergiant as large as the orbit of Mars, and slightly variable. ν is an attractive binocular double, mags 4·5 and 6·5, separation 41″. M4 and M80 are both globular clusters, M4 being the larger of the two and lying in the same binocular field as Antares itself. The open clusters M6 and M7 are among the finest in the sky. α (Antares): RA 16h 26m, Dec $-26\frac{1}{2}°$, mag about 1·1 (variable).

Scutum, the Shield
(Sct, Star Map 6)

A faint but distinctive group lying in a magnificent region of the Milky Way, containing the superb open cluster M11. α: RA 18h 32m, Dec $-8\frac{1}{2}°$, mag 3·8.

Serpens, the Serpent
(Ser, Star Map 6)

This constellation represents a snake being held by Ophiuchus, and is divided into head (Caput) and body (Cauda). It contains M5 (see chart), one of the largest and brightest globular clusters. α (Unukalhai): RA 15h 42m, Dec $+6\frac{1}{2}°$, mag 2·7.

Taurus, the Bull
(Tau, Star Map 3)
A prominent zodiacal group with more than its share of interesting objects. The bright, very scattered Hyades cluster, to the west of Aldebaran, looks in binoculars like a sprinkle of coloured jewels. The Pleiades, three times as distant at about 400 light-years, are much more compact, and contain very luminous white stars of a type that has burnt out and disappeared in the much older Hyades. At its distance of only 65 light-years, Aldebaran is twice as close as the Hyades. τ is an attractive binocular double, mags 4·5 and 8·5, distance 1′. The planetary M1 may just be spotted, about 1° NW of ζ; this is the Crab Nebula, the remains of the supernova of 1054. α (Aldebaran): RA 4h 33m, Dec $+16\frac{1}{2}°$, mag 0·9.

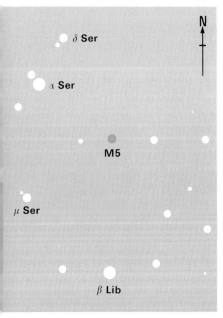

▼ **M5 in Serpens** rivals the Great Cluster in Hercules in brightness. It is 27,000 light-years away.

δ Ser

N

α Ser

M5

μ Ser

β Lib

Triangulum, the Triangle
(Tri, Star Map 2)
A small but fairly distinct group containing the interesting galaxy M33. At least as large as the Moon, it is excessively faint and difficult to see: a real challenge for the clearest of nights. Excellent dark adaption is essential. Let the binoculars slowly pass over the suspect area, and use averted vision, which means directing the gaze away from the region so that the image falls on the more sensitive margin of the retina. Only slightly further off than M31, it is a much smaller object, with about 1/40th of its mass. β: RA 2h 07m, Dec $+35°$, mag 3·0.

Triangulum Australe, the Southern Triangle
(TrA, Star Map 8)
A small high southern group on

▲ **Red Antares** in Scorpius glows above the summer horizon, as seen from southern England.

the edge of the Milky Way. NGC 6025 is a fine open cluster. α: RA 16h 43m, Dec −69°, mag 1·9.

Tucana, the Toucan
(Tuc, Star Map 8)
Containing the Small Magellanic Cloud, this constellation also includes the brilliant globular cluster 47 (NGC 104), which rivals ω Cen. β may be too difficult for binoculars: mags both 4·5, separation 27″. NGC 362 is another fine globular. α: RA 22h 15m, Dec −60½°, mag 2·8.

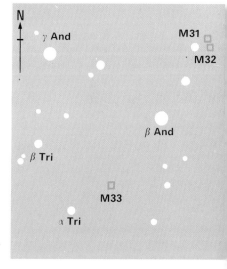

▲ **Elusive M33,** in Triangulum, can just be detected with binoculars in a transparent sky.

Ursa Major, the Great Bear
(UMa, Star Maps 1 and 5)
Sometimes known as the Plough or the Dipper, the outline of this constellation is known to most people, although it extends far beyond the seven bright stars. ζ (Mizar) is mentioned on page 52. The faint planetary M97 may be spotted with binoculars in a very dark sky. α (Dubhe): RA 11h 01m, Dec +62°, mag 1·8.

Ursa Minor, the Little Bear
(UMi, Star Map 1)
The north polar constellation, containing the Pole Star or Polaris. α (Polaris): RA 02h 15m, Dec +89°, mag 2·0.

Vela, the Sails
(Vel, Star Maps 4 and 8)
A bright Milky Way group. γ is a testing double (mags 1·8 and 4·2, distance 42″), just resolvable in good binoculars. The nearby cluster NGC 2547 can be seen with the naked eye. γ: RA 8h 08m, Dec –47°, mag 1·7.

Virgo, the Virgin
(Vir, Star Map 5)
A large zodiacal constellation, best known for the galaxies within its confines, though few are visible with binoculars. The chart on page 85 shows the positions of the brightest (M49, M58 and M87), as well as the fainter M60. α (Spica): RA 13h 23m, Dec –11°, mag 1·0.

Vulpecula, the Fox
(Vul, Star Map 7)
This is a very small Milky Way group, but it contains M27, one of the largest and brightest planetary nebulae in the sky (see chart). α: RA 19h 27m, Dec +24½°, mag 4·4.

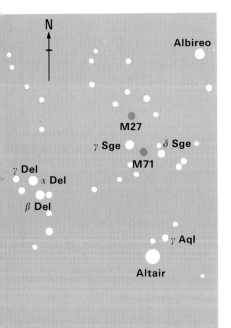

▶ **The large** planetary M27, the Dumb-bell nebula, is a bright binocular object, but is not easy to locate among the many faint stars in Vulpecula — see the chart (left). M71 in Sagitta is also shown.

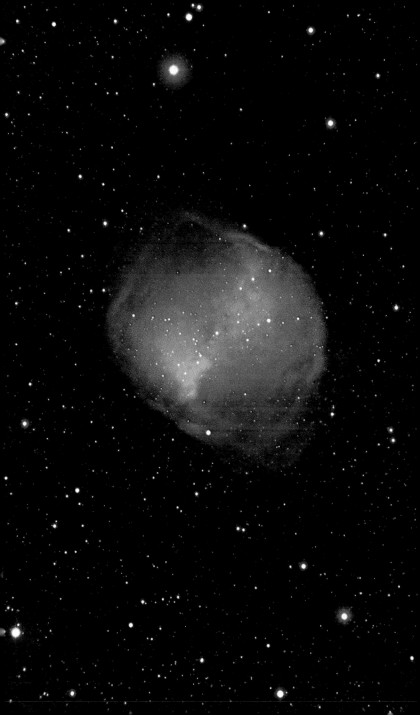

The Moon

Of all the objects in the sky, the Moon is the one that appears to undergo the most dramatic change, since it passes through a complete cycle of phases once a month. This change, though, is only one of movement and light. The lunar surface is inert, airless and dead; the last dramatic events happened there some 3000 million years ago, and they are still recorded, apparently fresh, on its crater-pocked face.

The Moon passes through phases because it shines only by reflecting sunlight. The Moon, like the Earth, is always half-lit by the Sun. As it goes through its phases night by night, different features of its surface are illuminated. A considerable amount of detail can be seen with a small telescope or binoculars, so it is not surprising that the Moon is the favourite object for anyone taking up astronomy as a hobby.

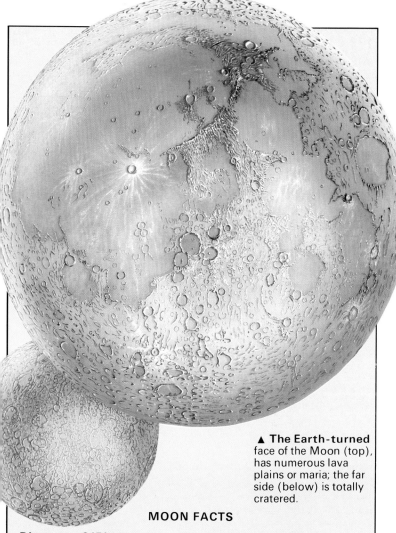

▲ **The Earth-turned**
face of the Moon (top),
has numerous lava
plains or maria; the far
side (below) is totally
cratered.

MOON FACTS

Diameter: 3476 kilometres
Mass: 0·012 X Earth
Density: 3·3 X water
Mean distance: 384,400
    kilometres
Minimum distance:
    356,400 kilometres
Maximum distance:
    406,700 kilometres

**True (sidereal) period of
    rotation:** 27·32 days
**Phase cycle (synodic
    period):** 29·53 days
**Inclination of orbit to
    ecliptic:** 5°
Axial inclination: $6\frac{1}{2}$°
Angular diameter: 29′ 21″ (min.);
    33′ 30″ (max.)

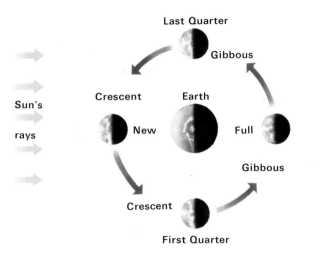

Last Quarter

Gibbous

Crescent

Earth

New

Full

Sun's

rays

Crescent

Gibbous

First Quarter

New Moon

First Quarter

Full Moon

Last Quarter

▼ **The Moon** shows phases because the Sun can illuminate only one hemisphere. It rotates once on its axis during the lunar month, therefore always keeping the same face towards the Earth.

First Quarter, 7 days   Full Moon, 14 days

The Moon's Cycle

It takes the Moon 29½ days to pass through its phases and this is known as the lunar month. During this time, the sunrise and sunset line, or *terminator,* slowly passes across the Earth-turned hemisphere: sunrise before Full, sunset after Full.

At New Moon, the dark hemisphere faces the Earth and cannot be seen, since it is very near the Sun in the sky. After two or three days, it has moved far enough east to be seen as a thin crescent in the evening sky; after seven days it forms a perfect half, known as First Quarter, lying about 90 degrees away from the Sun.

The following week sees it in the *gibbous* state, until at Full it is situated opposite the Sun in the sky, rising at sunset. After this, the phases pass in the reverse order, the Moon rising later and later in the night, until it disappears in the dawn sky.

Unless the line-up between the three bodies is perfect and an eclipse occurs, the Moon is invisible at New since it is hidden in the Sun's glare. It is interesting to try to spot the thin crescent setting in the west a couple of days after New or rising in the east a couple of days before New. Binoculars will be a help in locating the hair-thin line of light.

Last Quarter, 21 days   **Final Crescent, 27 days**

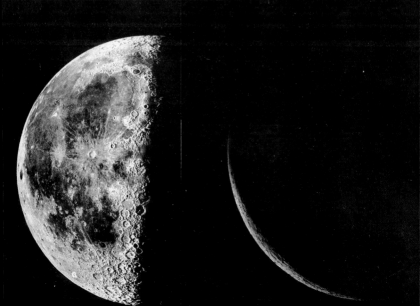

Lunar Eclipses

If the Moon moved exactly along the ecliptic (which could happen only if it revolved exactly in the plane of the Earth's orbit), the line-up with the Sun at New and Full would be perfect and there would always be an eclipse of the Sun and the Moon respectively at these times.

However, the Moon's orbit is inclined at five degrees to the ecliptic, so eclipses are rather rare. A lunar eclipse occurs when the Moon passes into the Earth's shadow. At a total eclipse it passes completely into the central shadow or umbra. An eclipse may last for several hours and may be total for well over an hour. But even when it is totally eclipsed, the Moon is always dimly visible because the Earth's atmosphere passes some light into the shadow, giving it a reddish-brown colour.

Total eclipses are particularly interesting, because some are much darker than others. This is due to the varying transparency of the atmosphere. Partial eclipses on the other hand are of little interest, since the dazzling uneclipsed portion of the surface blots out the delicate shadow colouring. The eclipse of June 1964 was so dark that the Moon could hardly be seen with the naked eye, but on most occasions the very bright and very dark features, such as maria and ray craters, are visible with binoculars.

▼ **An eclipse** of the Moon occurs when it passes through the long shadow cast by the Earth.

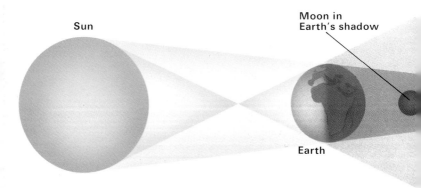

Sun

Moon in Earth's shadow

Earth

▲ **Four stages** in the progress of a lunar eclipse. It can take up to six hours for the Moon to pass completely through the Earth's shadow, and totality can last for $1\frac{3}{4}$ hours.

LUNAR ECLIPSES			
Date and Time (GMT) of mid-eclipse	Type	Duration (minutes)	Region of Visibility
1982 Jul 6 07.30	Total	102	New Zealand, Pacific, South America, Mexico
Dec 30 11.26	Total	66	Pacific, Australia, New Zealand, East Asia, USA
1983 Jun 25 08.25	Partial	—	South Pacific, New Zealand, South America
1985 May 4 19.57	Total	70	Africa, Indian Ocean, East Indies, Australia
Oct 28 17.43	Total	42	Asia, East Africa, West Australia

Occultations by the Moon

As the Moon (and a planet too, for that matter) passes along the ecliptic, it regularly moves in front of stars and blocks them from view. Such a phenomenon is called an *occultation*. Although occultations can be predicted for years in advance, there is always an uncertainty of a second or more in the exact instant at which the star vanishes or reappears.

Stars disappear at the eastern edge or limb, which before Full is invisible unless the Moon is a crescent and its dark side can be seen faintly illuminated by *Earthshine* – sunlight reflected on to it from the Earth. Dark-limb occultations are much easier to observe than bright-limb events, because the star is easily lost in the glare of the Moon. Planets are also occulted by the Moon and it can be very exciting to watch as the Moon gradually covers the planet's surface. The Moon moves surprisingly fast, and even the large disc of Jupiter is totally covered within about 80 seconds.

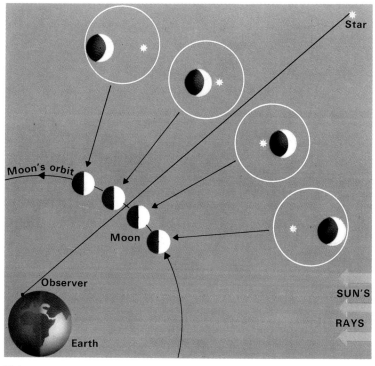

► **An occultation** of a star photographed with a 300-millimetre reflector. The first picture shows two stars. In the second one has been occultated by the Moon. The Moon is waxing gibbous, and the western limb is invisible; when it is waning, stars disappear at the bright limb.

◄ **By timing** the disappearance or reappearance of a star at an occultation, the Moon's position in its orbit can be accurately determined, since the star acts as a reference point in the sky.

Observing Occultations

First of all you need to know when occultations are expected: the information on page 183 tells you how to obtain occultation predictions.

Occultations of stars brighter than about 6th magnitude can be observed with a 60-millimetre aperture telescope, provided the Moon is not too near the Full and there is no haze. To find the exact time at which a star disappeared, you will also need a stop-watch and the use of a telephone. Using the stop-watch, press the button at the moment the star disappears; then dial the telephone number of the speaking clock. Stop the watch on a time signal – any one will do. Subtract the time shown on the watch from the time given by the telephone clock to obtain the instant of occultation.

Do pass your observations on to the local society. Much useful information about the motion of the Moon has been collected from occultation observations.

Amateur Observation

Thanks to the *Apollo* landings in 1969-1972, we know a great deal more about the Moon's surface and interior than could ever have been found out from the Earth. But the discoveries have been less sensational than those made about the planets, simply because we have such an excellent view of our satellite.

Even the naked eye will reveal the dark lava plains or *maria* ('seas'), and the bright uplands. Brilliant individual patches such as the white deposit around

Copernicus (Lunar Chart 3) can also be discerned.

It is interesting to compare the naked-eye view with a chart, and to discover how much small detail can be seen. The bright crater Kepler (Chart 3) is more difficult than Copernicus, but the dark central sea Mare Vaporum (Chart 2) is harder still.

Remember that the charts show an inverted view of the Moon, to agree with the upside-down view of most astronomical telescopes.

Binoculars or a low-power telescope show the main lunar

(1)

(2)

(5)

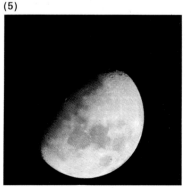

(6)

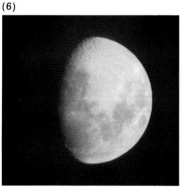

features perfectly well, but for an adequate view of the craters and mountain chains in their true majesty, an aperture of 75 millimetres or more and a magnification of at least 100 is necessary. The view then is breathtaking.

The Lunar Charts

For convenience, the Earth-turned surface shown on the charts that follow has been divided into four quadrants. Objects in the 1st and 2nd quadrants are best observed between New and First Quarter (local sunrise), and again between Full and Last Quarter (sunset).

The 3rd and 4th quadrants can be seen under sunrise illumination between First Quarter and Full, and under sunset conditions between Last Quarter and New Moon.

▼ **These photographs** were taken by an amateur astronomer using a camera attached to a 60-millimetre refractor and with no special guiding. The age of the Moon in days is as follows: (1) 2·8; (2) 3·8; (3) 5·8; (4) 7·6; (5) 9·0; (6) 10·0; (7) 11·7; (8) 14·0.

(3)

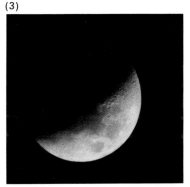

(4)

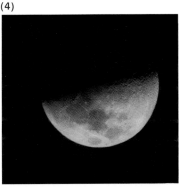

(7)

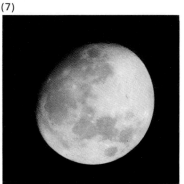

(8)

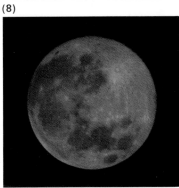

Lunar Chart 1
South-East Quadrant

To aid identification, the objects on these charts are shown as they would appear under sunset illumination, when they lie near the terminator.

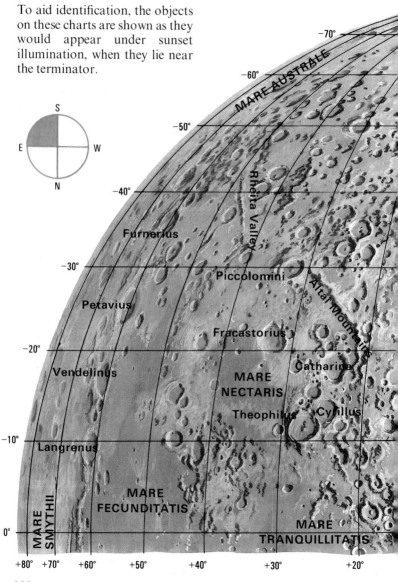

108

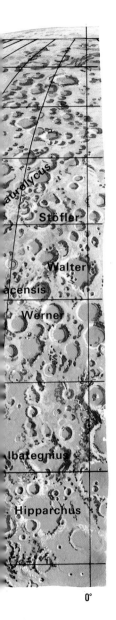

Rheita Valley: Best seen at four days old, this 160-kilometre long fault is never very easy to make out because of its tumbled neighbourhood.

Altai Mountains: One of the more prominent mountains ranges in this quadrant. They look rather like the border of a wide scooped-out valley, and rise in places to 4000 metres.

Theophilus: This crater, 100 kilometres across, has overlapped its neighbour Cyrillus, proving that it was formed at a later date.

Fracastorius: A good example of an old crater whose wall has been partly melted and broken down by fresh lava from the newly formed Mare Nectaris.

Petavius: This makes an imposing sight with Vendelinus, Langrenus and Furnerius along the terminator of the three-day-old Moon. Measuring 160 kilometres across, it has a wide valley running from the central mountain to the south-west wall.

Hipparchus: This and its neighbour **Albategnius** are both about 150 kilometres across and bear all the signs of extreme age, having been peppered by meteoritic impacts long after they were first formed by gigantic collisions early in the solar system's history.

Stöfler: Prominent because of the intrusions on its east wall, this 80-kilometre crater can be recognized quite easily even in the chaotic terrain that characterizes the southern lunar highlands.

0°

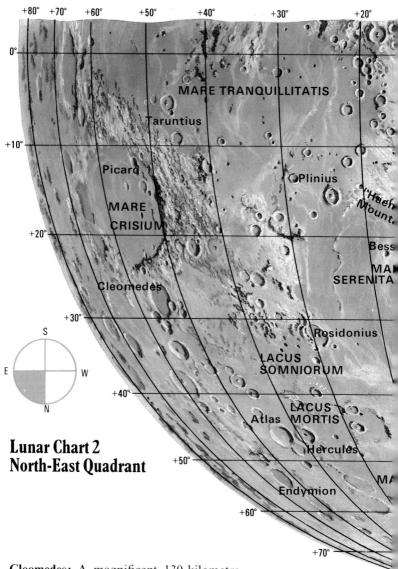

+80° +70° +60° +50° +40° +30° +20°

0°

MARE TRANQUILLITATIS

Taruntius

+10°

Picard

Plinius

Haeh Mount.

MARE CRISIUM

Bess

+20°

MA SERENITA

Cleomedes

+30°

Posidonius

LACUS SOMNIORUM

+40°

LACUS MORTIS

Atlas

Hercules

Lunar Chart 2 North-East Quadrant

+50°

MA

Endymion

+60°

+70°

S
E — W
N

Cleomedes: A magnificent 130-kilometre crater, spectacular about two days after Full, when it and the Mare Crisium make a superb sight.

110

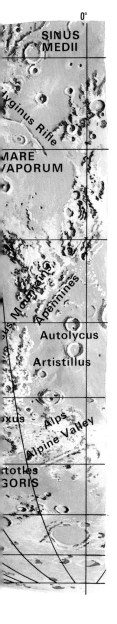

SINUS MEDII

yginus Rille

MARE
VAPORUM

Montes

Apennines

Autolycus

Artistillus

xus

Alps

Alpine Valley

ttotles

GORIS

0°

Mare Crisium: One of the smaller seas, only 500 kilometres across, but very prominent. Its distance from the East limb changes markedly throughout the month, due to the Moon's *libration* – a slight swinging on its axis due to its changing orbital velocity getting out of step with its constant axial spin. Libration allows an observer to see a small part of the 'invisible' hemisphere.

Mare Serenitatis: A partly mountain-bordered sea, best seen at about five days old, when the surface is seen to be crossed by numerous ridges, as if it had wrinkled like a skin over the still-warm interior.

Bessel: A small crater, only about 20 kilometres across, but easily spotted on the relatively crater-free surface of the Mare Serenitatis. A long bright ray crosses both it and the mare.

Endymion: Compare the dark floor of this 130-kilometre crater with its lighter neigh-bours Atlas and Hercules.

Caucasus Mountains: These and the Apennines are among the highest on the Moon, rising in places to over 6000 metres.

Hyginus Rille: A long valley, visible at First Quarter as a thread-like line if a small telescope is used.

Alpine Valley: The Alps are not the highest mountain range on the Moon, but they contain the most extraordinary valley, 130 kilometres long and over 10 kilometres wide. At First Quarter it is visible with a 60-millimetre telescope as a sharp cut.

111

Lunar Chart 3
North-West Quadrant

Plato: A beautiful dark-floored crater 100 kilometres across. Evidently the interior was flooded by re-melting.

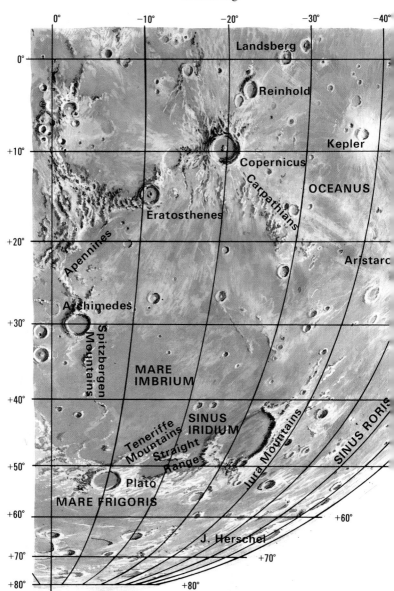

Landsberg

Reinhold

Kepler

Copernicus

Carpathians

OCEANUS

Eratosthenes

Aristarc

Apennines

Archimedes

Spitzbergen Mountains

MARE IMBRIUM

Teneriffe Mountains

SINUS IRIDIUM

Straight Range

Jura Mountains

SINUS RORIS

Plato

MARE FRIGORIS

J. Herschel

0°   −10°   −20°   −30°   −40°

0°
+10°
+20°
+30°
+40°
+50°
+60°
+70°
+80°

+60°
+70°
+80°

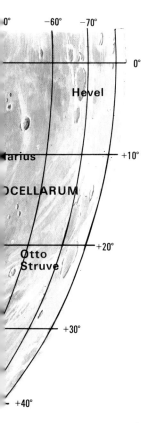

0°   −60°   −70°

0°

Hevel

arius   +10°

OCELLARUM

Otto
Struve   +20°

+30°

+40°

S

E   W

N

Archimedes: This fine 90-kilometre crater, with its smaller companions Autolycus and Aristillus, makes a fine sight just after First Quarter. The floors of all three have been re-melted, presumably by the same action that produced the Mare Imbrium.

Mare Imbrium: Perhaps the most beautiful of all the seas, as well as being one of the largest, measuring about 800 kilometres across.

Straight Range: A small but obvious mountain group that stands up conspicuously on the Mare surface near Plato.

Sinus Iridum: Probably once a huge crater 250 kilometres across, but now one wall has been reduced by flooding, leaving a magnificent bay with peaks (the Jura Mountains) rising to 6000 metres. When the Moon is about nine and a half days old they can be seen with the naked eye, jutting over the terminator.

Copernicus: One of the youngest large craters on the Moon. Ninety kilometres across, it shows all the characteristic features: terraced walls, surrounding ridges and pits from the impact, a central peak, and the white rays caused by glassy molten fragments that soared across hundreds of kilometres of ground.

Aristarchus: A small crater only 48 kilometres across, the brightest spot on the lunar surface. It can often be seen near local midnight, under Earthshine conditions (that is, when the Moon is a four-day-old crescent); it also shows up well during a total lunar eclipse.

113

Lunar Chart 4
South-West Quadrant

Clavius: One of the largest lunar craters, 230 kilometres across, with a chain of more recently-formed craters inside it.

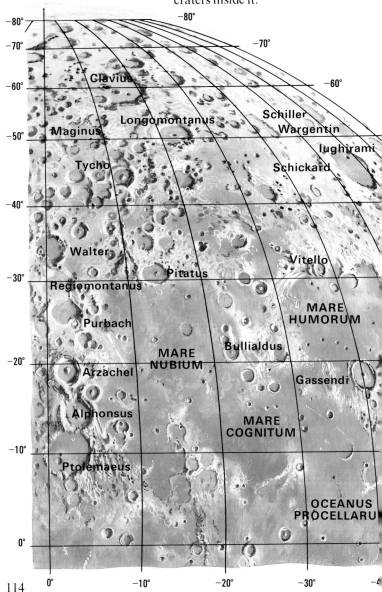

S

E —|— W

N

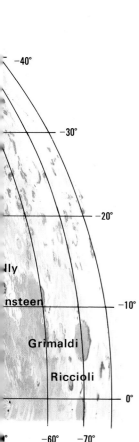

−40°

−30°

−20°

lly

nsteen   −10°

Grimaldi

Riccioli

0°

−60°   −70°

Ptolemaeus: This ruined 145 kilometre crater, together with Alphonsus and Arzachel, are imposing just after the First Quarter; the western walls are sunlit long before their bases are fully illuminated.

Oceanus Procellarum: The largest sea, but without mountainous borders, and much of it is rather featureless with a small aperture. There are many drowned rings where ancient craters have been submerged.

Grimaldi: A very dark crater near the west limb, and always obvious because of its tint. The slightly smaller neighbour Riccioli (160 kilometres across) is much harder to make out, except just before Full, when it lies on the terminator.

Gassendi: A beautiful crater with a flooded floor 90 kilometres across. Amateurs study Gassendi closely, because faint temporary red colorations have been reported in this area.

Mare Humorum: One of the better-defined small seas. A fine sight on the 11-day-old Moon, when tiny craterlets and ridges seam the surface.

Schickard: A dark-floored 210-kilometre crater, ruined by later impacts. Note the curious object Wargentin to the south-west: a unique plateau, apparently caused when the original crater became filled with lava.

Tycho: A well-formed crater 90 kilometres across, obviously a relatively recent, well-preserved formation like Copernicus, and the centre of the most extensive ray system on the Moon.

SUN

The Planets

The planets have been mysterious objects ever since the ancient astronomers puzzled about their movements across the celestial sphere. We now understand their wanderings; but as 'worlds' there are still many questions unanswered about them.

The planets can be divided into two groups: the small or *terrestrial* planets, Mercury, Venus, Earth, Mars and possibly Pluto, whose mass is in their solid globes; and the four *giant* planets, Jupiter, Saturn, Uranus and Neptune, which consist predominantly of gas and have little or no rocky material.

▲ **The planets** with their orbits drawn to scale.
From left to right: Mercury, Venus, Earth, Mars,
Jupiter, Saturn, Neptune, Uranus and Pluto.

The planets, satellites, comets and scraps of debris that make up the
Sun's family were all formed, together with our star, inside a huge
cloud that probably consisted of 90 per cent hydrogen, 9 per cent
helium, and just traces of the other 90 naturally-occurring elements.

Hydrogen and helium have very fast-moving atoms that require a
strong gravitational pull to restrain them. The Sun, and the giant
planets – Jupiter, Saturn, Uranus and Neptune – were able to do so,
and consist mostly of hydrogen. The smaller planets, like the Earth,
lost their hydrogen very early on and are dense, rocky bodies.

117

Planet Fragments

Many other bodies also condensed from the original solar nebula. Some became satellites of the planets, while other smaller relics are the thousands of minor planets or *asteroids,* orbiting mainly in the region between Mars and Jupiter. Innumerable millions of dust-sized fragments, *meteoroids,* circle the Sun invisibly unless they burn up as *meteors* in the Earth's atmosphere. Finally there are the curious

▶ **On the Earth's surface,** living cells based on the carbon atom have developed into countless varieties of life forms. Life began over 3400 million years ago, the date of the earliest known fossils.

SOLAR SYSTEM FACTS

Planet	Distance from Sun (millions of km) Min	Mean	Max	Diameter (km)	Length of day	Length of year	No. of satellites
Mercury	46	58	70	4850	176d	88d	0
Venus	107½	108	109	12,104	2760d	225d	0
Earth	147	149½	152	12,756	24h	365d	1
Mars	206½	228	249	6790	24h 37m	687d	2
Jupiter	741	778½	815½	142,600	9h 50m	11·9y	14
Saturn	1347	1427	1507	120,200	10h 14m	29·5y	17
Uranus	2735	2870	3004	52,000?	24h?	84y	5
Neptune	4456	4497	4537	48,000?	22h?	165y	2
Pluto	4425	5900	7357	3000?	6d 9h	248y	1

comets, icy and vaporous, that brighten to visibility only as they sweep near the Sun during their lonely passage.

A question which intrigues every astronomer is whether there is life on the other planets in the solar system. Certainly the conditions on most other planets make this unlikely. Their temperatures, for example, are too extreme, fluctuating from severe cold to searing heat, for the plant and animal life of Earth.

▲ **Spacecraft** have photographed many planets and satellites. The results suggest that every solid surface in the solar system, including the Earth's, was bombarded by small interplanetary bodies in its early history. Venus (top), Mercury (below).

▶ **The surface** of Mars as seen from a *Viking* spacecraft.

Observing the Planets

The planets are always fascinating to watch. Due to their orbital motion around the Sun, they are never in exactly the same position on the celestial sphere from night to night. However, different planets move in different ways.

Inferior and Superior Planets

Mercury and Venus, which are called the *inferior planets*, are a special case. Their orbits lie inside that of the Earth, and so they are always in the neighbourhood of the Sun. Pretending, for simplicity, that the Earth is stationary, the lower left diagram shows how these planets move. First they pass from invisibility on the far side of the Sun, a position called *superior conjunction,* to eastern elongation, visible in the western sky after sunset. Then they go through inferior conjunction

again, usually invisible, and out to western or morning elongation. During this time they pass through phases like the Moon, appearing Full near superior conjunction and as a thin crescent near inferior conjunction.

The other planets behave as in the right-hand diagram. When closest to the Earth they are opposite the Sun in the sky, like the Full Moon – a position known as opposition; near conjunction they are invisible, behind the Sun. Only Mars sometimes shows a slight phase, near quadrature. These planets,

SUPERIOR PLANET

INFERIOR PLANET

120

▲ **Binoculars** or a small telescope should reveal Jupiter's four large moons when they are near elongation from the planet.

▲ **A powerful astronomical** telescope will enable you to see a mass of fine, constantly-changing detail on Jupiter's disc.

from Mars to Pluto, are known as the *superior planets*.

The planets from Mercury to Saturn are visible with the naked eye; Uranus and Neptune can be seen with binoculars. Only Pluto is invisible without a proper astronomical telescope of about 250 millimetres aperture. Binoculars will reveal the crescent phase of Venus and up to four moons of Jupiter, as well as one of Saturn's moons. Binoculars are also a great help in finding shy Mercury during its fleeting appearances.

▶ **When Mars,** or any other superior planet, comes to opposition, the Earth's greater orbital speed makes it appear to backtrack or retrograde for a few weeks.

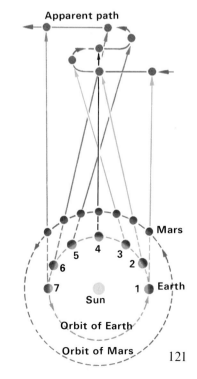

121

Mercury ☿

Mercury always lies within 28 degrees of the Sun, and from temperate latitudes it can be seen only in twilight, when it is near elongation. From the northern hemisphere, the best time for hunting it is from January to April as an evening object, and from July to October as a morning object. In the southern hemisphere, these times are reversed.

Finding Mercury

Look about ten degrees above the Sun's place about three-quarters of an hour before sunrise or after sunset. Times of forthcoming elongations are given in the Sky Diary. But start looking for the planet about ten days before the time given. Mercury is white but if you are looking for it in the evening, the sky often colours it pink.

Mercury shows such a tiny disc that a magnification of 250 shows it only as large as the naked-eye Moon. It was not until *Mariner* 10 visited it in 1974-1975 that much was found out about it.

We now know that it is crater-ridden like the Moon. Its surface, alternately baking and freezing, is airless and dead.

MERCURY FACTS

Surface temperature:
350°C/−170°C
Gravity: 0·38 X Earth
Density: 5·4 X water
Atmosphere: none
Apparent diameter: 9″ (inferior conjunction); 6″ (superior conjunction)
Interval between inferior conjunctions: 116 days
Maximum magnitude: −1·4

▼ **Mercury** as seen through a 250-millimetre reflector on three different evenings.

▲ **Mercury** has an exceptionally large metal core, which makes it the densest planet in the solar system after the Earth.

Core
Mantle
Crust

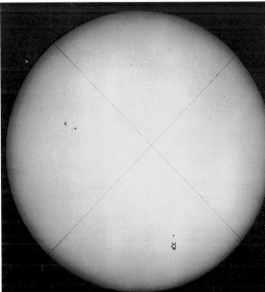

▶ **Mercury** sometimes transits the Sun at inferior conjunction. This photograph was taken on May 8, 1970. The next transit will be on November 13, 1986.

Venus ♀

Like Mercury, Venus is an inferior planet, swinging into and out of the Sun's rays. It is much easier to see, however, and near elongation it can be made out with the naked eye in the daylight sky.

Binocular Observation

Between elongation and inferior conjunction, the crescent phase can be seen using firmly-mounted binoculars. Unfortunately, not even a powerful telescope can reveal more than faint shadings in the planet's cloudy atmosphere. Most of our knowledge of Venus has come from space probes.

Rapid Movement

Venus is usually lost from view for about four months around the time of superior conjunction. But it sweeps through inferior conjunction so quickly that it reappears in the morning sky only a matter of days after vanishing in the evening twilight. Venus is called the Morning or Evening Star according to when it is visible.

VENUS FACTS

Surface temperature:
480°C
Gravity: 0·9 X Earth
Density: 5·2 X water
Atmosphere: mainly carbon monoxide
Atmospheric pressure:
91 X the Earth's
Apparent diameter: 62″ (inferior conjunction): 10″ (superior conjunction)
Interval between inferior conjunctions: 584 days
Maximum magnitude: −4·4

▼ **These drawings,** made with apertures of 90 millimetres and 150 millimetres, show faint cloud shadings.

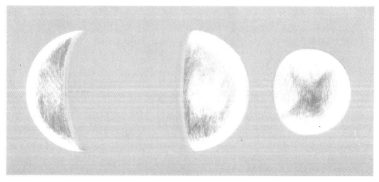

▲ **Venus'** internal structure is probably similar to that of the Earth.

Core

Mantle

Crust

▼ **The phases** of Venus.

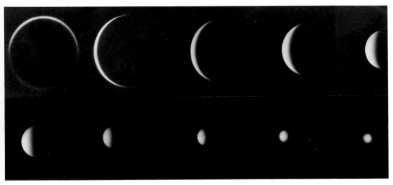

Mars ♂

Mars is the only planet in the solar system that gives us a reasonable view of its true surface. But because it is so rarely well-placed for observation, it has puzzled rather than informed amateur and professional astronomers alike.

Mars comes to opposition every 26 months or so, but its orbit is so eccentric that some apparitions bring the planet much closer than do others. In 1971 it came within 56 million kilometres of the Earth, but in 1980 its opposition distance was about 100 million kilometres.

Observation Hints

This means that observing Mars is a frustrating business. Really favourable oppositions occur at intervals of about 17 years; the next will be in 1986. Mars also approaches and recedes so rapidly that useful work can be carried out only for a few weeks. When it does appear, you can try to follow its rapid path with binoculars. With telescopes of 60 millimetre aperture you can look out for dark markings and the polar caps. Try following a dark marking for about an hour. It

---

MARS FACTS

Surface temperature:
  −20°C/−200°C
Gravity: 0·38 X Earth
Density: 3·95 X water
Atmosphere: very thin, mainly carbon dioxide
Apparent diameter:
  14″-25″ (opposition); 3″ (conjunction)
Interval between oppositions: 780 days
Maximum magnitude: −2·5

▶ **A photograph** of Mars taken from Earth.

▼ **Three drawings** of Mars made by amateur astronomers.

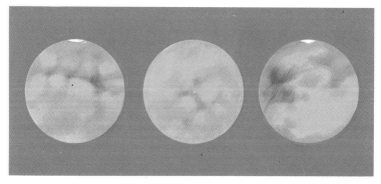

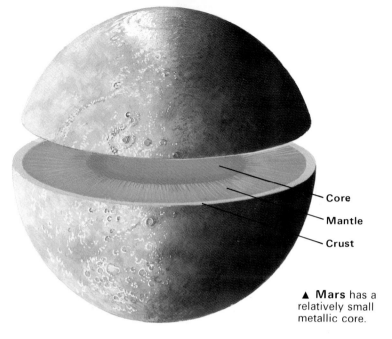

Core

Mantle

Crust

▲ **Mars** has a relatively small metallic core.

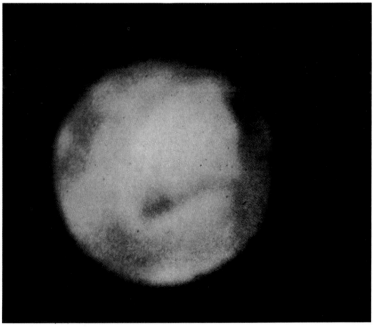

will move with the planet's rotation from east to west (right to left in the telescope).

The Deceptive Planet

In 1896, the American observer, Percival Lowell, built an observatory in Arizona dedicated to the study of Mars. He believed that he had discovered dozens of artificial waterways or canals on its surface, and that the planet was inhabited by intelligent beings. Unfortunately, the thin straight lines observed by Lowell and others do not exist. The tiny Martian disc had deceived them into imagining these features.

Mars' Atmosphere

The atmosphere of Mars is very thin – less than a hundredth the density of our own – and consists mainly of carbon dioxide. But fierce winds can blow the dust into huge clouds, visible from the Earth by the dark features they obliterate; even the polar caps have sometimes disappeared. The strongest dust storms happen when Mars is near perihelion, at its closest to the Sun.

White clouds are also seen, usually near the planet's limb, where they can be almost as bright as a polar cap. Like the caps, they consist of ice.

65°	120°	60°	0° NORTH
Diacria	Arcadia	Acidalium Mare	Ismenius Lacus
30°			
Amazonis	Nix Olympica Tharsis	Lunae Palus	Oxia Palus	Arabia
0° WEST			
Memnonia	Phoenicis Lacus	Coprates	Margaritifer Sinus	Sabaeus Sinus
-30°			
Phaethontis	Thaumasia	Argyre	Noachis
-65°			SOUTH
180°	120°	60°	0°

Life on Mars

Today, even after the sensational *Viking* landings in 1976, we still have not finally answered the question whether there is life on Mars. If there is, though, it can only be on the bacterial level.

The two sites that were examined gave no definite evidence of living organisms. But the surface of Mars is so varied – parts are cratered desert, like the Moon, while others exhibit relatively recent volcanic features – that some other regions may be more hospitable. Unfortunately, no further probes are likely for many years.

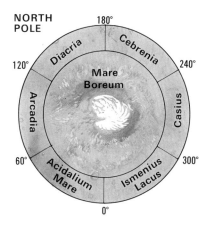

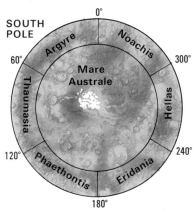

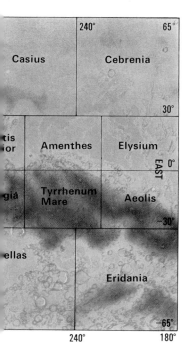

◄ **These maps** show the surface of Mars as photographed by space probes. The dark areas visible from Earth are shaded in. Two of the most obvious features are the dark Syrtis Major and the light Hellas. The huge volcano on the Tharsis-Amazonis ridge, Nix Olympica, is some 30 kilometres high with a crater at the top 65 kilometres across.

Jupiter ♃

Jupiter is an ideal planet for the amateur to observe. Its disc is always large (even a magnification of about forty makes it look as big as the Moon), and there are four bright satellites. In order outward from the planet these are Io, Europa, Ganymede and Callisto. They can all be seen using good binoculars.

With an aperture of 60 millimetres and a magnification of one hundred or so, details are visible on the 'surface' (really the upper cloud layer). The belts show wisps and irregularities, and the famous Great Red Spot can be seen when it is prominent, although it sometimes fades from view.

Jupiter spins so rapidly that its disc is noticeably flattened at the poles. A watch of between five and ten minutes will show that its markings are moving from east to west across the disc, due to its rotation. The drawings below show the amount of detail that can be recorded with normal amateur telescopes.

Observing the Satellites
You can also observe the four satellites as they move around the planet. Try timing their orbital paths by seeing how long they take to return to a position chosen. Try to make out their shadows as they move across Jupiter's disc.

JUPITER FACTS

Temperature at cloud tops: −150°C
Gravity: 2·7 X Earth
Density: 1·3 X water
Overall composition: hydrogen, helium
Apparent diameter: 47″ (opposition); 32″ (conjunction)
Intervals between oppositions: 399 days
Maximum magnitude: −2·5

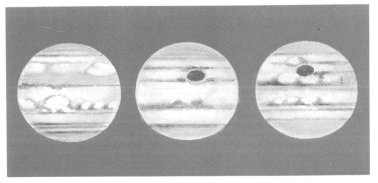

▲ **Jupiter** from a *Voyager* spacecraft.

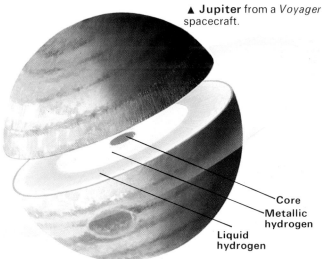

Core
Metallic hydrogen
Liquid hydrogen

◄ **Drawings** of Jupiter made with an aperture of 100 millimetres.

▲ **Jupiter** is mainly gaseous or liquid, with a tiny core.

131

Saturn ♄

Saturn used to be known as the planet with the rings, but we now know that it shares this distinction with Jupiter and Uranus. However, the rings around these other planets are so faint that they have never been seen by eye, whereas Saturn's are one of the glories of the night sky.

Earth-based telescopes can distinguish three main rings, but the two *Voyager* spacecraft have revealed hundreds of strands only a few kilometres wide; some are concentric while others are braided together.

The rings lie in the plane of Saturn's equator. Its axis is tilted, so that during its long 'year' we see the rings at different angles.

Markings and Moons

Saturn's globe is even more flattened than Jupiter's. Its rotation period is slightly longer, but it has very little rigidity – its average density is less than that of water. The belts are much fainter than those of Jupiter, and noticeable spots are rare.

◄ **Two close-up** views from *Voyager* of Saturn and its rings, showing the numerous ringlets.

► **During Saturn's** 29½-year revolution, each pole and ring face is inclined in turn towards the Earth and Sun.

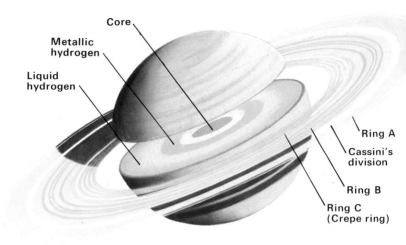

Core
Metallic hydrogen
Liquid hydrogen
Ring A
Cassini's division
Ring B
Ring C (Crepe ring)

One of the brightest markings ever seen on Saturn was discovered in 1933 by the famous comedy actor and amateur astronomer, Will Hay, using a 150-millimetre reflector.

At least 17 satellites have been discovered – eight by spacecraft – ranging from Titan, 5200 kilometres across, to objects smaller than many asteroids.

Observing Saturn

Saturn's rings can be seen with a magnitude of around 50 and, with higher magnification, the dark Cassini division between them may be observed. With good binoculars, Titan can be made out and, in small telescopes, other moons can be identified. It is difficult to make out any markings on Saturn's disc as it is so small.

▲ **Saturn is** probably built on a very similar pattern to Jupiter.

SATURN FACTS

Ring system –
Outer diameter: 272,300
Inner diameter: 149,300
Temperature at cloud tops: −180°C
Gravity: 1·2 X Earth
Density: 0·7 X water
Overall composition: hydrogen, helium
Apparent diameter of globe: $19\frac{1}{2}''$ (opposition); $16''$ (conjunction)
Interval between oppositions: 378 days
Maximum magnitude: −0·3

134

Uranus ⛢, Neptune ♆ and Pluto ♇

▲ **A drawing** of the distant planets (not to scale). From right to left: Uranus, Neptune and Pluto.

Our knowledge of the three outermost planets is scanty. The rotation periods and diameters of Uranus and Neptune are uncertain, while Pluto is too small to show a disc.

Uranus is just visible to the naked eye, but it was not noticed until 1781. The discovery was made by William Herschel who came across the new planet in a survey of the sky with a 150-millimetre reflector. Neptune and Pluto were found after deliberate searches by professional astronomers: Neptune on the basis of predictions by Adams and Leverrier in 1846 and Pluto by Tombaugh in 1930.

Rings and Satellites

Both Uranus and Neptune seem to be built on the giant-planet pattern. They each have small solid cores with thick layers of hydrogen, helium and methane (a compound of hydrogen and carbon) forming an icy, slushy covering. Uranus is particularly strange: it has faint rings, discovered in 1977 when they occulted and dimmed a star, and its axis is tilted so far over that, in the course of its long year, each pole comes to point almost directly at the Sun.

While Uranus has five satellites, Neptune has only two: one, Triton, is larger than the Moon, and the other, tiny Nereid, moves in a most eccentric orbit. At times it swings six times nearer than at others.

Little Pluto, too, has a curious satellite called Charon, about half its own diameter. The orbit of Pluto is so eccentric that between 1979 and 1999 it comes closer to the Sun than Neptune. It is quite possible that small planets beyond Pluto remain to be discovered.

Observation Hints

Both Uranus and Neptune can be made out with binoculars if their position in the sky is known. But you will need to refer to astronomical yearbooks (see page 183) to know in which part of the sky to start hunting for them.

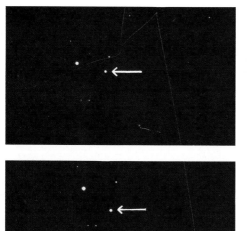

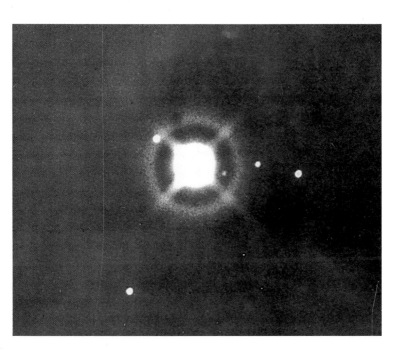

▲ **Uranus** with its five known satellites. The 'steering-wheel' appearance is a photographic effect.

◄ **Neptune has** only two known moons: Triton (near the planet) and Nereid, whose orbit carries it from about $1\frac{1}{2}$ million to 10 million kilometres from Neptune.

◄ **Pluto is too small** to be seen as a planetary disc. Its identity is revealed on these two photographs, (far left), taken on different nights, by its altered position when compared with nearby stars. Pluto can be seen as a faint speck with an aperture of about 250 millimetres.

137

Solar System Debris

The planets of the solar system formed when tiny particles inside the solar nebula began adhering together. In some cases the process was halted before it had gone very far, and small bodies a few kilometres across were all that resulted. Larger bodies probably collided, and broke up. The result of these misfortunes is the zone of asteroids or minor planets, which is concentrated between the orbits of Mars and Jupiter.

Asteroids

The largest asteroid, Ceres, is 1000 kilometres across, but most of the 2200 that have been discovered are far smaller. Vesta, 540 kilometres across, can sometimes reach 6th magnitude, and about a dozen are within the range of binoculars. A few asteroids have orbits that take them near the Earth.

▼ **Most known asteroids** keep to the zone between the orbits of Mars and Jupiter. Icarus, Adonis, Eros and Chiron are among the few with eccentric orbits.

▶ **This brilliant fireball** was accidentally caught by an amateur photographing the stars.

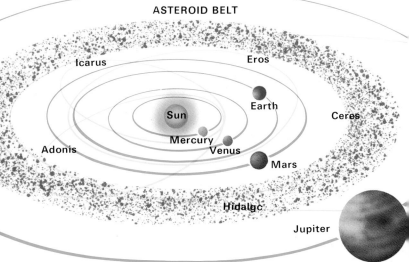

ASTEROID BELT

Icarus · Eros · Earth · Sun · Ceres · Mercury · Venus · Adonis · Mars · Hidalgo · Jupiter

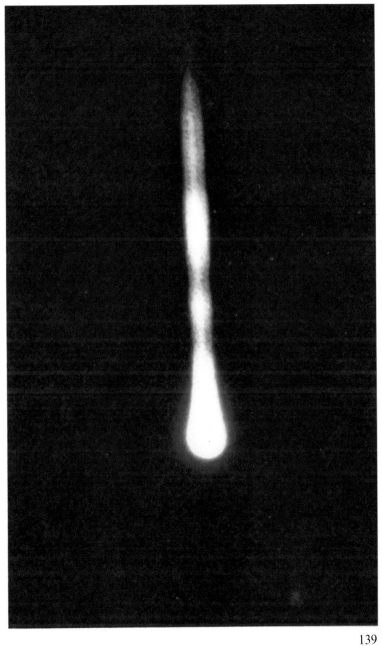

Observing Asteroids

An asteroid can be distinguished from a star only by its nightly motion, or by using a good star atlas. If its approximate position in the sky is known, an accurate drawing of the field will show one 'star' to have shifted. Asteroids will also show up as short streaks in long-exposure photographs taken anywhere near the ecliptic.

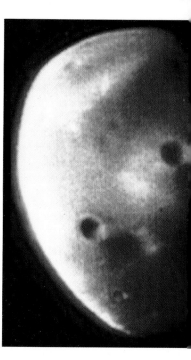

► **The surface of Mars'** tiny moon Deimos was photographed in 1976 by a *Viking* orbiter. In appearance it probably resembles many asteroids.

▼ **This photograph** shows the tracks of numerous meteors that flashed across the sky during the brilliant Leonid display of November 1966. Such brilliant displays are very rare.

Meteors

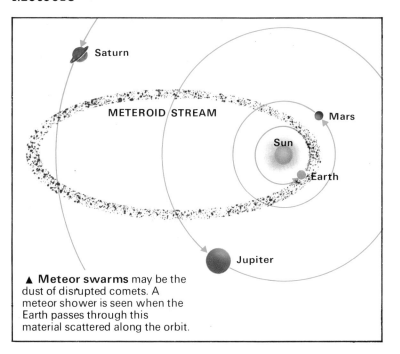

▲ **Meteor swarms** may be the dust of disrupted comets. A meteor shower is seen when the Earth passes through this material scattered along the orbit.

Meteroids – fragments of solid matter not more than a few centimetres across – pervade the solar system. Some orbit the Sun independently, while others travel in swarms that have scattered themselves along a particular orbit.

The Earth moves in its orbit at about 30 kilometres per second, and when it encounters a meteroid, the relative velocity of the two bodies can be anything up to about 60 kilometres per second if it is a head-on collision. At these higher speeds the meteroid is vaporized in the atmosphere, leaving the streak of light we see as a shooting star or *meteor*.

Individual objects produce *sporadic* meteors, which are seen at a constant rate throughout the year. When the Earth passes through a swarm, however, a so-called meteor *shower* occurs. This recurs every year when the Earth returns to its intersection with the swarm.

The Earth sometimes collides with much larger bodies weighing several kilogrammes. These leave a brilliant streak that lights up the whole landscape, and may even reach the ground as a *meteorite*.

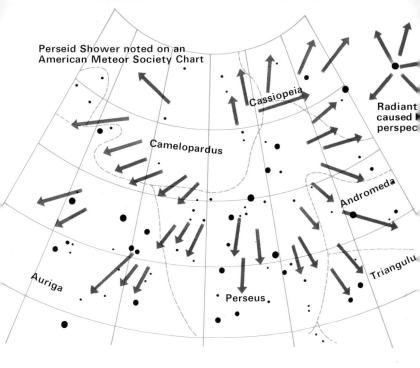

Perseid Shower noted on an American Meteor Society Chart

Cassiopeia

Camelopardus

Radiant caused perspec

Andromeda

Auriga

Triangulu

Perseus

Observing Meteors

Sporadic meteors can be seen on any clear night, being most frequent and brighter in the early morning. Meteor showers, however, occur only at certain times of the year – see the table opposite. Bright moonlight can drown the fainter meteors, so observations are affected by the Moon's phases.

Observing a meteor shower is even more fun, since a group of up to six observers is best. Four people are allotted a quarter of the lower sky each, one looks directly overhead, and the sixth writes down details of the meteors seen.

The most useful data to record are the time (to within half a minute), magnitude, speed (fast, medium or slow), colour, and whether a train (a faint after-trail) was left. It is also interesting to work out how many meteors per hour were seen (the Hourly Rate or HR).

Experienced observers can try noting the meteor paths in relation to the stars. Shower meteors all appear to come from one place in the sky – the *radiant*. The constellation in which the radiant lies give the meteor shower its name. Often observers prepare in advance charts, like the one above, on which meteor paths can be marked.

142

METEOR SHOWERS

Some meteor showers have very sharp maxima that lasts only a few hours, and these times vary from year to year because of leap-year adjustments. The Sky Diary gives the best times at which to observe the Quadrantids, April Lyrids, Perseids and Geminids, all of which have brief maxima. The two Aquarid showers are best observed from the southern hemisphere.

Shower	Noticeable activity	Maximum activity	Maximum HR (approx.)
Quadrantids	Jan 1–6	Jan 3–4	50
April Lyrids	Apr 19–24	Apr 22	10
η Aquarids	May 1–8	May 5	10
δ Aquarids	Jul 15 – Aug 15	Jul 27	25
Perseids	Jul 25 – Aug 18	Aug 12	50
Orionids	Oct 16–26	Oct 20	20
Taurids	Oct 20 – Nov 30	Nov 8	8
Leonids	Nov 15–19	Nov 17	6?
Geminids	Dec 7–15	Dec 14	50

▶ **The Barringer** meteorite crater in Arizona may be 50,000 years old. It is 1300 metres in diameter. Below is a meteorite showing the pitted surface caused by vaporization during its fall through the Earth's atmosphere.

Comets

The popular idea of a comet is of a long-tailed object gleaming in the sky. But fewer than one in a hundred achieves this distinction. Most are so faint that they can be seen only as a faint smudge, even with a powerful telescope, while countless more must pass through space undetected.

Comets, like asteroids, are probably original members of the solar system. They are small bodies, much more crumbly than rock, and contain frozen water and other compounds. In most cases their orbits are very eccentric, taking them from the region of the inner planets to Pluto or beyond.

When they approach the Sun, the frozen material vaporizes, releasing the crumbly material it has been cementing together. The gas and dust form a *coma,* which is all that astronomers usually see of a comet. If it passes very near the Sun, solar radiation may sweep the material back into a long tail.

◀ **Comet West, 1976,** was one of the brightest comets of the century. It is photographed here over Kitt Peak National Observatory.

▼ **Comet Kohoutek** discovered in 1973 by Lubos Kohoutek at Hamburg Observatory.

Observing Comets

Amateurs are still very successful in hunting for new comets. This involves sweeping regularly over the night sky with an aperture of between about 100 and 200 millimetres, with a low magnification, for example × 30.

Since comets are brightest when near the Sun, the western sky after dusk and the eastern sky before dawn are the most promising areas to search. A very dark transparent sky gives the best chance of making a discovery; but it may take some time. A few comets such as Kohoutek's are discovered by chance but most are found after years of patient hunting.

▼ **Halley's Comet** recedes beyond the orbit of Neptune at aphelion, but it is now heading for the vicinity of the Sun.

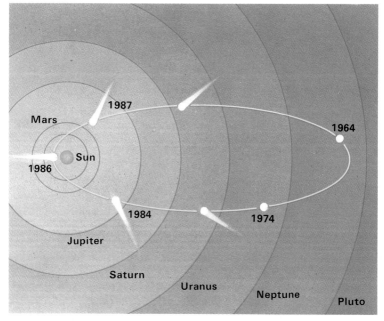

146

▲ **Halley's Comet** and Venus photographed in 1910 in South Africa.

▲ **A drawing** of Halley's Comet over Jerusalem in 66 AD. From Lubienietski's *Historia Universalis Omulum Cometarum,* 1966.

SOME IMPORTANT COMETS			
Name	Orbital period (years)	Last seen	Remarks
Encke	3·3	1980	Sometimes visible with binoculars; the shortest period known.
Biela	6·6	1852	Seen to break up. A meteor shower occurred when the Earth next passed near its position.
Schwassmann-Wachmann	16·1	—	Always more distant than Jupiter, and comes to opposition every year. Occasional outbursts.
Halley	76·0	1910	Next perihelion passage 1986.
Daylight comet	750?	1882	Probably the brightest comet of modern times.
Donati	1900?	1858	Famous for its curved tail, 40° long.
Daylight comet	4 million?	1910	Probably the brightest comet of the 20th century.

Atmospheric Astronomy

If the Earth had no atmosphere (ignoring the fact that life would then be impossible), the astronomer's task would be much simpler. The sky would always be transparent, and free from clouds, and stars near the horizon would shine as brightly as those overhead.

But some features of the night sky would be absent. *Aurorae* occur in the upper atmosphere, as do the rare *noctilucent* clouds. Other features, which we include under 'atmospheric astronomy' although they really occur out in space, would be seen more clearly, particularly the *Zodiacal Light, Zodiacal Band,* and the *Gegenschein* or *Counterglow.*

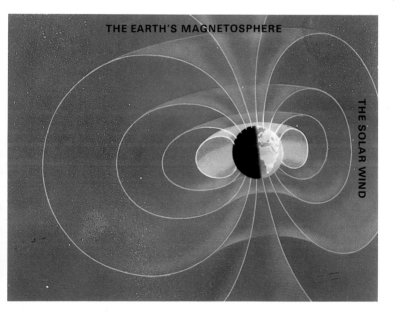

THE EARTH'S MAGNETOSPHERE

THE SOLAR WIND

◄ **A bright aurora** is a memorable sight. Displays are most common around the time of sunspot maximum, when the Sun sometimes emits fierce bursts of radiation that makes atoms in the upper atmosphere glow.

▲ **The Earth** has the most powerful magnetic field of all the terrestrial planets. Electrically-charged atomic particles from the Sun are trapped in different shells, some positive (protons) and some negative (electrons).

149

Aurorae

Aurorae are connected with solar activity. Near ground level, the atmosphere consists of nitrogen, oxygen and other elements existing as stable molecules. But at a height of about 50 kilometres, the air is so thin that the molecules are easily split into atoms, and the atoms themselves broken down, by the Sun's energy.

When the Sun's surface is particularly active, it radiates atomic particles that energize these atoms, and the energy is given off as light. The result is an auroral display. Aurorae occur mostly in the upper atmosphere near the Earth's poles, where the solar particles find an easier entry down through the magnetic field. Aurorae can last for hours and sometimes all through the night.

Observation Notes

A faint aurora usually takes the form of a diffuse glow above the northern horizon (the southern horizon, in the southern hemisphere). It may develop rays extending up towards the zenith, and green and red tints may appear.

If you suspect an auroral display, make regular (for example five-minute) records of its height and extent along the horizon. The best latitude for auroral observations is about 60 degrees north or south, but strong displays can sometimes be seen close to the equator.

Zodiacal Light

Zodiacal Light is caused when sunlight is reflected by interplanetary dust. It takes the form of a cone extending along the ecliptic. Since it is so faint, it can be seen only during a critical time between the beginning or end of twilight and its own rising or setting.

In higher latitudes it makes a low angle with the horizon, and is difficult to see. In latitudes of about 35 degrees and less, it is bright enough to drown the fainter stars, and is known as the 'false dawn'.

Look for it in the west after dusk in the spring, and in the east just before dawn in the autumn (southern observers should reverse these times). A dark country sky is essential, and your eyes must be thoroughly dark-adapted if the elusive cone is to be seen.

Gegenschein

The Gegenschein is also caused by interplanetary dust, but it lies opposite the Sun in the sky, and appears as a very dim patch several degrees across. It is fainter than the Zodiacal Light or the Milky Way. The best time to observe it is around midnight. In the northern hemisphere look for it at the beginning of November, when it lies in Aries (Star Map 2 and 3). The best chance for southern observers is in early February, when it lies in Capricornus (Star Map 7).

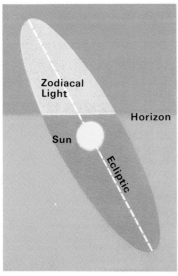

▲ **The Zodiacal Light** is a permanent phenomenon of the night sky. It is caused by sunlight reflected off countless dust particles in the plane of the solar system, and therefore has a very elongated shape. Because it is so faint, photographs of Zodiacal Light are very rare. Above is an artist's impression of the phenomenon seen in Scotland.

◄ **A diagram** to show how the Zodiacal Light appears in relation to the Sun.

151

PHOTOGRAPHIC MAGNITUDES

The Milky Way and Beyond

All the stars visible in the night sky belong to the Milky Way, our Galaxy. For a long time astronomers had no clear idea of what the Galaxy was like because the view from inside the star-system is a very poor one. Then they realized that other galaxies can also be seen and a picture of our own Galaxy emerged.

The Galaxy is a pinwheel-shaped spiral, like the Andromeda Galaxy two million light-years away. It is one of a group of about 20 that form a big cluster, the Local Group. Most of the galaxies in the Local Group are elliptical-shaped dwarfs, much smaller than the Milky Way. Our cluster measures about two and a half million light-years across – not large compared with some groups, such as the Coma cluster which contains thousands of galaxies. Astronomers can make out the brighter individual stars in these neighbour galaxies, and they are similar to those in the Milky Way.

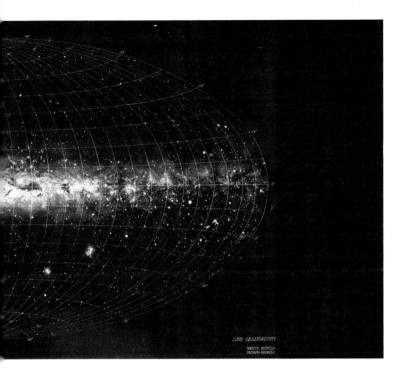

LUND OBSERVATORY

THE GALAXY

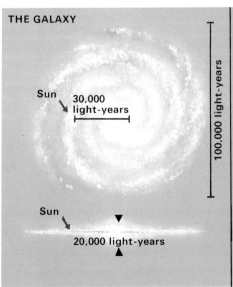

Sun
30,000
light-years

100,000 light-years

Sun

20,000 light-years

▲ **This is how the
Galaxy** would appear
if viewed from out in
space. The position of
the Sun (see diagram
left), 30,000 light-
years from the centre,
means that we see
many more stars when
looking towards the
nucleus (in the
direction of Sagittarius).
The Galaxy contains
upwards of 100,000
million stars, but at
least half of its
material is in the form
of non-luminous
interstellar dust. This
picture does not show
the halo of globular
clusters.

Galaxies in Space

Galaxies are scattered throughout space as far as telescopes can reach, but it is not easy to measure their distances. For many years, astronomers doubted whether any galaxies outside the Milky Way could be detected at all. For example, they thought M31 in Andromeda was a local star cluster.

But in the 1920s it was realised that M31 must be remote, since Cepheid variables were discovered in it, looking much fainter than the furthest Cepheids found in the Galaxy. Later on, novae were also found, and their assumed absolute magnitudes helped to place M31 at a distance of 2,200,000 light-years away.

Unfortunately, individual stars such as Cepheids and novae cannot be made out in very remote galaxies and other methods must be used to measure their distance. One way is through the red-shift relationship (see page 160). Another is to make assumptions about the absolute magnitude of the whole galaxy. (The absolute magnitude of M31, for example, is about –21, equivalent in brightness to 25,000 million stars as luminous as the Sun.)

The difficulty with this method is that galaxies vary greatly in size and brightness. Some of the dwarf galaxies in the Local Group have absolute magnitudes as low as –9, which is not much brighter than a single highly-luminous supergiant star! Others are much more luminous than the Milky Way.

It seems that, although galaxies may differ greatly from one to another, the types of stars they contain can all be found on the Milky Way-based Hertzsprung-Russell diagram. Star formation is a standard process throughout the universe.

▲ **M82 in Ursa Major,** an irregular galaxy about $8\frac{1}{3}$ million light-years away. The Seyfert galaxy (below) has an unusually bright nucleus.

► **The Whirlpool Galaxy** M51 was the first spiral discovered. M31 in Andromeda (below) is a bright nearby counterpart of the Milky Way.

Classes of Galaxies

Tens of thousands of galaxies have been photographed in some detail, and most fall into the following distinct classes: spiral (normal and barred), elliptical (like a spiral's nucleus, without arms), and irregular.

The star populations in these classes differ considerably. Irregular galaxies such as the Magellanic Clouds contain many young stars and nebulae (from which more new stars can form). Spirals have a mixture of young and old stars. In the arms there are young, white giant stars, Sun-like stars, dying white dwarfs and also gas and dust. In the nucleus are mostly old red giants.

Elliptical galaxies are the commonest type, and include some of the largest objects known. The stars are mainly red giants. There are also some very small examples, containing dim red stars: the faintest galaxies in the Local Group are of the elliptical class, as are the two bright satellite galaxies of M31 in Andromeda. The lack of dust and gas means that no new stars can be forming in these particular star systems.

▶ **The Magellanic Clouds** are visible to observers in the southern hemisphere.

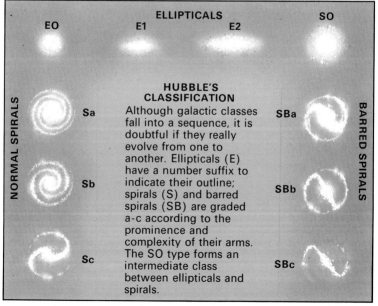

NORMAL SPIRALS

ELLIPTICALS

EO   E1   E2   SO

BARRED SPIRALS

Sa   SBa

Sb   SBb

Sc   SBc

HUBBLE'S CLASSIFICATION

Although galactic classes fall into a sequence, it is doubtful if they really evolve from one to another. Ellipticals (E) have a number suffix to indicate their outline; spirals (S) and barred spirals (SB) are graded a–c according to the prominence and complexity of their arms. The SO type forms an intermediate class between ellipticals and spirals.

Observing Galaxies

Part of the fun of observing galaxies lies in the chase itself, for most are difficult to find and will be a challenge for the most careful observer.

The sky notes on pages 80-97 give a good sample of the brighter galaxies, but only one in the whole sky, M31 in Andromeda, can be seen with the naked eye. If it happens to be well placed, spend some minutes examining it with binoculars or a small telescope. It may seem just a faint haze, but ask yourself the following questions, and others like them, and try to answer them accurately.

What shape is it? What direction is the long axis? What is the size in degrees or minutes of arc? How does the brightness change from centre to edge? Is the brightest part in the centre? Are there any dark or bright lanes, or condensations? Are any stars very near it, or projected on it? Is any colour detectable? If seen, what does the satellite galaxy M32 look like?

When you look at *anything* in the sky, ask appropriate questions like these. Your eyes will respond, and your observing will improve.

▼ **The Big Bang.** The Universe is thought to have begun with a massive explosion (1). Hydrogen began to form (2) and galaxies to condense (3). More highly developed galaxies (4) continue to move apart away from the explosion.

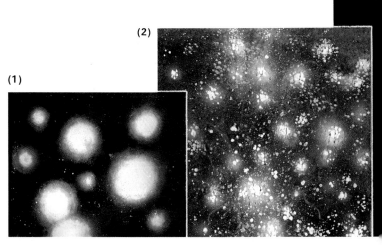

The Origin of the Universe

Most astronomers accept the idea that the universe is expanding from the *Big Bang*. Galaxies are flying apart, and if their tracks are run backwards, they must have been very close together about 20,000 million years ago.

Physicists have tried to analyse what must have happened to the material that gave rise to what we see now as the universe. If all the matter present now was present then, the 'primeval atom' must have been a mass of atomic particles at a temperature of *millions* of millions of degrees. Once it started expanding, however, the temperature would have dropped dramatically. Eventually, the chaos of particles must have started arranging itself into the elements we see today, hydrogen atoms being the simplest and by far the most common.

Evidence for the Big Bang theory comes from the discovery of a faint radiation pervading space. The only satisfactory explanation is that it represents the last traces of the primeval atom explosion, the flash of unbelievable heat that has almost faded away.

(4)

Red-Shift

Light travels as tiny pulses moving through space at 300,000 kilo-metres a second. The colour of the light depends on the distance between the pulses, the *wavelength*. If a light source such as a star or galaxy is moving rapidly away from an observer, the wavelength is 'stretched'. A longer wavelength shifts the lines in the object's spectrum towards the red end, since red light has a longer wavelength than blue. All the clusters of galaxies in the night sky except those in the Local Group show red-shift and so indicate the expansion of the universe.

Those galaxies whose distances can be measured fairly accurately follow a law which says that the amount of red-shift of the spectral lines is proportional to their distance. In other words, 'the farther, the faster'. If this law is correct, the red-shifts of very distant galaxies give a clue to their distance. The most remote objects so far observed are about 10,000 million light-years away.

The actual value of the so-called 'Hubble constant' is still in doubt because of the difficulty of making observations, but it is somewhere around an increase in expansion rate of 30 kilometres per second for every million light-years of distance.

▼ **If all the galaxies** are flying apart, why is there no point from which they are all receding? Since the universe has no edge, the situation resembles the relative motion of spots on a balloon. As it is blown up, the stretching rubber carries the spots apart, but it is impossible to find one spot that is not moving.

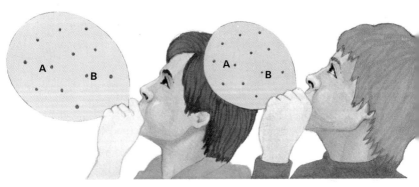

▲ **Radio telescopes** are sensitive to long-wave electromagnetic radiation. This one at Raisting in West Germany receives signals from artificial satellites while others can detect radiation from remote galaxies.

▼ **A shining body** emits electromagnetic radiation, which is energy pulses or waves. If it is moving at a high speed, its motion increases or decreases the wavelength as recorded by an observer.

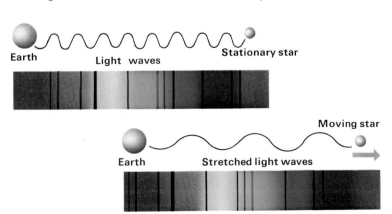

Pulsars and Quasars

Pulsars were detected in 1967 as faint regular pulses of radio signals, so brief and artificial-looking that they were once thought to be artificial messages.

Further work proved that they come from interstellar objects, some of which are spinning at a rate of 30 times per second. The only possible known candidate is a *neutron star,* the collapsed core of a supernova, in which the star's remaining material is compressed into a body only a few kilometres across. A pinhead of neutron-star material would weigh as much as a battleship.

This was proved when the faint star at the centre of the Crab Nebula, the famous supernova remnant (page 56), was found to be pulsing at a high speed. The intense magnetic field of the star may be responsible for forcing its radiation into a narrow beam like that of a lighthouse.

Quasars

Quasars, or quasi-stellar objects, are the most powerful known energy-sources in the universe. They appear to be much smaller than ordinary galaxies (although there is considerable uncertainty about their size), but emit hundreds of times as much radiation. They are hot, blue objects, and at least some are variable in brightness.

▼ **A pulsar** emits a narrow beam of radiation, probably caused by the fierce magnetic field curving the light into a stream. When the beam is pointed away from the Earth (1 and 2) the pulsar is invisible; at 3 the astronomer will observe a burst of radio emission or visible light. Most pulsars are radio objects.

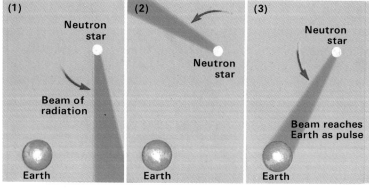

(1) Neutron star — Beam of radiation — Earth

(2) Neutron star — Earth

(3) Neutron star — Beam reaches Earth as pulse — Earth

▶ These two photographs reveal the pulsar at the centre of the Crab Nebula. It is alternately bright (top) and invisible (below) 30 times every second.

▼ Quasar 3C 273, about 600 million light-years away, is one of the closest and brightest known. (Negative print.)

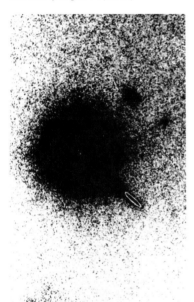

Large Red-Shifts

All quasars have very large red-shifts, which suggests that they are very far away; we are seeing them as they were thousands of millions of years ago, when the universe was younger than it is today. One significant puzzle is why there are no nearby quasars, as there are nearby galaxies. Perhaps they are short-lived objects that were formed early in the history of the universe.

Objects such as quasars and pulsars count among the most significant astronomical discoveries of the century. Just as the recent *Voyager* probes have made physicists puzzle over the nature of Saturn's rings, so the energy-production of quasars and the formation of pulsars have made scientists re-examine what they know about the nature of the physical world.

Simple Astrophotography

Astrophotography can be done without a telescope at all. Excellent star photographs can be taken using fast colour film such as Ektachrome 200 or 400, and any miniature camera with a lens opening of between f/2 and f/4. Black-and-white film, for example Tri-X or HP5, means that the processing can be done cheaply at home and the results seen the same evening

The star photographs on this page were all taken in this way. The trails below right are the result of leaving the camera shutter open for about ten minutes; the reddish trail near the centre is Alpha Herculis. If the shutter is left open for a minute or less, the stars hardly have time to trail, and a picture like the one of Cygnus opposite is the result. The photograph of Coma Berenices was taken with a guided camera and a ten-minute exposure.

▲ **A 35-millimetre camera** ready to take a time exposure of the sky. The cable release minimizes any shake as the shutter opens.

▶ **Como Berenices,** (top right).

▶ **Cygnus** (centre) photographed with a one-minute unguided exposure. On the original slide stars as faint as magnitude 8 could be seen. Deneb (α Cygni) is at the left.

▶ **A 10-minute exposure** reveals the brighter naked eye stars as curved streaks, due to the rotation of the Earth. The reddish star to the left of centre is α Herculis; to the left of that is ι Ophiuchi.

Astrophotography with a Telescope

To take close-up photographs, the camera must be fitted to the telescope so that the image formed by the objective lens or mirror is focused sharply on the film. Some telescopes have a special attachment, or you may have to be ingenious and make do with cardboard and sticky tape. The camera lens must be removed first.

Exposures must be shorter than about a second, or the Earth's rotation will make the image blur. This does not matter with solar photography, where a slow film, a neutral density filter, and fast shutter speed are essential.

To obtain a larger image, use an eyepiece between the telescope's image and the camera. This amplifies the image and allows more detail to be recorded, but you will need some help from a more comprehensive book on practical astronomy (see page 183). The photographs on the opposite page were taken by Peter Crabtree in this way, using the 150-millimetre reflector shown above right.

▶ **A medium-sized** refracting telescope with a camera attached for direct photography of the sky. The pictures of the Moon on pages 106 and 107 were taken in this way, with exposures of about one second.

▶ **This equatorially-mounted** reflecting telescope of 150 millimetres aperture is shown set up for photographing the night sky. It was used to take the photograph below.

▼ **To take successful** pictures of the Sun, a very dense filter is necessary. This photograph (below right) shows the solar disc around the time of maximum activity in 1980.

▼ **The Moon's surface** offers plenty of challenge. Here the crater Tycho stands out amidst the southern lunar highlands.

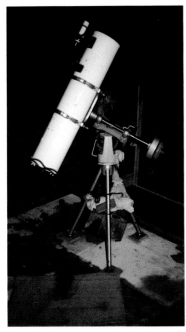

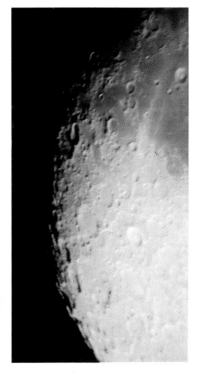

Making a Sundial

An equatorial sundial consists of a gnomon parallel with the Earth's axis which casts the shadow, and a plate, whose plane is at right angles to the gnomon, marked with lines to indicate the hours.

Cut out a right-angled triangular style (1), with one angle equal to your latitude. Set it on a base as in (2), and mount the dowel gnomon along its edge with adhesive.

To make the plate, draw a circle on a piece of thin card, divide it into 24 segments of 15 degrees each (3), and fold the card so that one semi-circle is on each side (4).

Mount the folded card against the face of the style so that it is

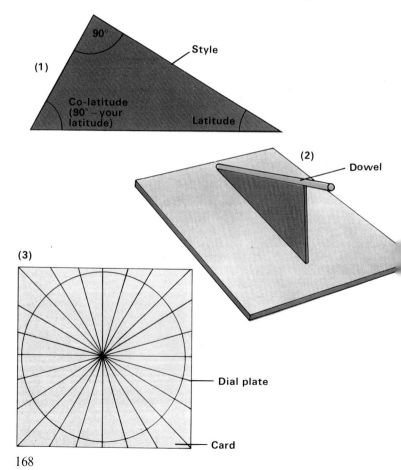

(1)

90°

Style

Co-latitude (90° – your latitude)

Latitude

(2)

Dowel

(3)

Dial plate

Card

168

accurately square-on to the gnomon (5), and inscribe each line with the hour it represents (the middle line will be 12 noon), as in (6).

To set the sundial up, place it on a horizontal surface and twist it until the shadow of the gnomon falls on the centre line at local noon.

Local noon will be *early* if you live *east* of your country's standard meridian, *late* if you live *west* of it. The difference is equal to four minutes for every degree of longitude. All sundials show local time. Don't forget to allow for the equation of time (see page 33).

During the six winter months, the Sun is south of the celestial equator and shines on the hour lines on the underside of the plate.

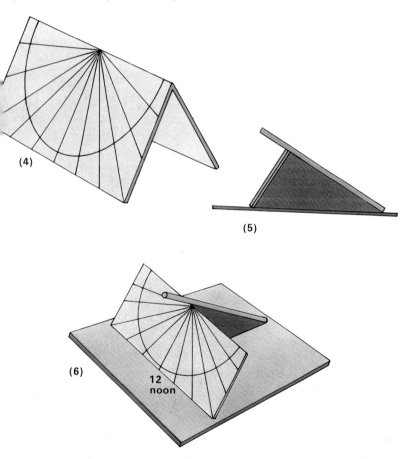

(4)

(5)

(6)

12 noon

169

Building a Planetarium

Trace the long rectangular map, and whichever polar map you need (see pages 172-174), and then prick the stars through into thin card (1). Use a darning needle (one millimetre thick) for the bright stars (blue), a pin for the medium stars (red), and prick a tiny hole for the others (black).

Cut out a thick disc 86 millimetres across. Pass a piece of bamboo through the centre and fix a torch bulb in its holder to one end (2) and (3).

Staple or glue the long star map into a cylinder, and fit one end to the thick disc (4). Stick the circular polar map to the other end, making sure the stars are in correct alignment (5). If there are cracks at the joins, cover them with aluminium foil (6).

Switch on the bulb and re-pierce the holes so that the light can be seen shining clearly. Stars near the edge of the cylinder must be pierced at an angle in a direction towards the bulb. Secure the bamboo loosely, so that it can be rotated, to a block of wood whose lower angle is equal to your latitude, and the stars will rise and set on the walls of a small room (7).

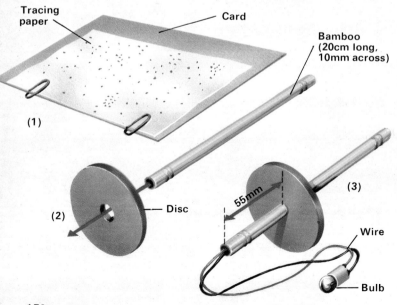

Tracing paper

Card

Bamboo (20cm long, 10mm across)

(1)

Disc

(2)

55mm

(3)

Wire

Bulb

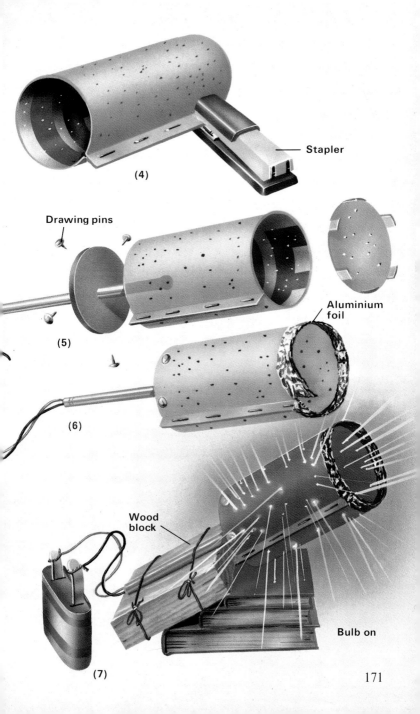

(4) Stapler

Drawing pins

(5)

Aluminium foil

(6)

Wood block

Bulb on

(7)

171

Trace these maps on to a single sheet of paper so that the upper edge of this page (12h RA) coincides with the lower edge on the opposite page. The dashed line indicates how the pierced card should be cut before it is curved into a cylinder.

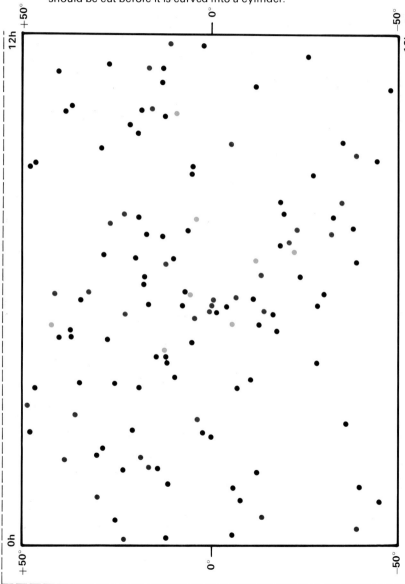

FOLD AND STAPLE HERE

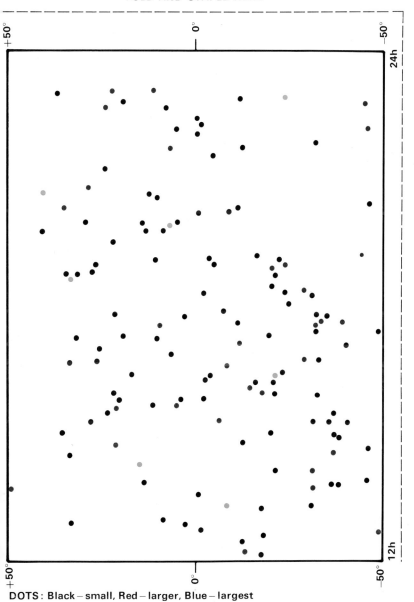

DOTS: Black – small, Red – larger, Blue – largest

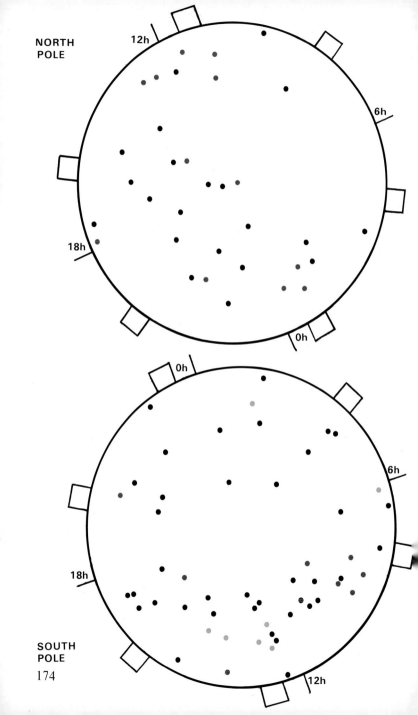

NORTH POLE

12h

6h

18h

0h

SOUTH POLE

174

0h

6h

18h

12h

A Sky Diary

This Diary records just some of the interesting events in the sky over the next five years. All times are GMT. (If you want to observe other minima of Algol the times can be calculated by knowing that its exact cycle of brightness change takes 68h 49m.) More detailed predictions can be found in astronomical yearbooks such as the B.A.A. Handbook (see page 183 for details).

1982

April	1	(Thu)	Venus at morning elongation.
April	8	(Thu)	Saturn at opposition in Virgo, magnitude 0·5.
April	21	(Wed)	April Lyrid meteor shower at maximum at 22.00. Moonlight will not interfere.
April	25	(Sun)	Jupiter at opposition on the border of Virgo and Libra, magnitude –2·0.
May	8	(Sat)	Mercury at evening elongation (21°), favourable for northern observers.
July	6	(Tue)	Total lunar eclipse, 07.30 (see page 103).
Aug	12	(Thu)	Perseid meteor shower at maximum at 10.00. Old Moon will interfere towards dawn.
Sep	6	(Mon)	Mercury at evening elongation (27°), favourable for southern observers.
Oct	17	(Sun)	Mercury at morning elongation (18°), favourable for northern observers.
Nov	4	(Thu)	Venus passes superior conjunction and becomes an evening object.
Dec	14	(Tue)	Geminid meteor shower at maximum at 07.00, Moonlight will not interfere.
Dec	16	(Thu)	Algol at minimum, 21.00.
Dec	30	(Thu)	Total lunar eclipse, 11.26.

1983

Jan	4	(Tue)	Quadrantid meteor shower at maximum at 04.00. The Last Quarter Moon will interfere before dawn.
Jan	8	(Sat)	Algol at minimum, 19.35.
Feb	8	(Tue)	Mercury at morning elongation ($25\frac{1}{2}°$), favourable for southern observers.

March (end of)			Maximum of long period variable χ in Cygnus to magnitude 4 or magnitude 5.
April	21	(Thu)	Mercury at evening elongation (20°), favourable for northern observers. Saturn at opposition in Virgo, magnitude 0·4.
April	22	(Fri)	April Lyrid meteor shower at maximum at 12.00. The Moon, approaching Full, will interfere badly with observations.
May	27	(Fri)	Jupiter at opposition on the border of Scorpius and Ophiuchus, magnitude −2·1.
June	11	(Sat)	Total solar eclipse, maximum duration 5m 11s (see page 40).
June	16	(Thu)	Venus at evening elongation.
June	25	(Sat)	Partial lunar eclipse, maximum phase 34% (see page 103).
Aug	12	(Fri)	Perseid meteor shower at maximum at midnight (24.00). Moonlight will interfere during the early evening only.
Aug	19	(Fri)	Mercury at evening elongation ($27\frac{1}{2}°$), favourable for southern observers.
Aug	25	(Thu)	Venus passes inferior conjunction to become a morning object.
Oct	1	(Sat)	Mercury at morning elongation (18°), favourable for northern observers.
Nov	4	(Fri)	Venus at morning elongation.
Dec	4	(Sun)	Annular solar eclipse (see page 40).
Dec	14	(Wed)	Geminid meteor shower at maximum, 22·00. Moonlight will interfere during the early evening only.
Dec	18	(Sun)	Algol at minimum, 21.20.

1984

Jan	4	(Wed)	Quadrantid meteor shower at maximum, 10.00. Moonlight will not interfere.
Jan	10	(Tue)	Algol at minimum, 20.00.
Jan	22	(Sun)	Mercury at morning elongation (24°), favourable for southern observers.
April	3	(Tue)	Mercury at evening elongation (19°), favourable for northern observers.
April	21	(Sat)	April Lyrid meteor shower at maximum, 19.00. Moonlight will interfere towards dawn.
May	3	(Thu)	Saturn at opposition in Libra, magnitude 0·3.

May	11	(Fri)	Mars at opposition in Libra, diameter 17·6″, magnitude –1·7.
May	30	(Wed)	Annular solar eclipse (see page 40).
June	15	(Fri)	Venus passes superior conjunction to become an evening object.
June	29	(Fri)	Jupiter at opposition in Sagittarius, magnitude –2·2.
July	31	(Tue)	Mercury at evening elongation (27°), favourable for southern observers.
Aug	12	(Sun)	Perseid meteor shower at maximum, 08.00. Full Moon will interfere.
Sep	14	(Fri)	Mercury at morning elongation (18°), favourable for northern observers.
Nov 22-23		(Thu/Fri)	Total solar eclipse, maximum duration 1m 59s (see page 40).
Dec	14	(Fri)	Geminid meteor shower at maximum, 05·00. Moonlight will interfere towards dawn.
Dec	19	(Wed)	Algol at minimum, 21.50.

1985

Jan	3	(Thu)	Mercury at morning elongation (23°), favourable for southern observers. Quadrantid meteor shower at maximum, 16.00. Moonlight will interfere until shortly before dawn.
Jan	11	(Fri)	Algol at minimum, 20.20.
Jan	22	(Tue)	Venus at evening elongation.
Mar	17	(Sun)	Mercury at evening elongation ($18\frac{1}{2}°$), favourable for northern observers.
April	3	(Wed)	Venus passes inferior conjunction to become a morning object.
April	21	(Sun)	April Lyrid meteor shower at maximum at midnight (24.00). Moonlight will not interfere.
May	4	(Sat)	Total lunar eclipse, 19.57 (see page 103).
May	15	(Wed)	Saturn at opposition in Libra, magnitude 0·2.
Jun	13	(Thu)	Venus at morning elongation.
Aug	4	(Sun)	Jupiter at opposition in Capricornus, magnitude –2·3.
Aug	12	(Mon)	Perseid meteor shower at maximum, 12·00. Moonlight will interfere towards dawn.
Aug	28	(Wed)	Mercury at morning elongation (18°), favourable for southern observers.
Oct	28	(Mon)	Total lunar eclipse, 17.43 (see page 103), favourable for southern observers.

Nov	8	(Fri)	Mercury at evening elongation (23°),
Nov	12	(Tue)	Total solar eclipse, maximum duration 1m 55s (see page 40).
Dec	14	(Sat)	Geminid meteor shower at maximum, 10.00. Moonlight will not interfere.
Dec	21	(Sat)	Algol at minimum, 22.20.

1986

Jan	3	(Fri)	Quadrantid meteor shower at maximum, 23.00. Last Quarter Moon will rise late.
Jan	13	(Mon)	Algol at minimum, 20.50.
Feb	28	(Fri)	Mercury at evening elongation (18°), favourable for northern observers.
April	13	(Sun)	Mercury at morning elongation (28°), favourable for southern observers.
April	22	(Tue)	April Lyrid meteor shower at maximum, 08.00. The almost Full Moon will interfere.
April	24	(Thu)	Total lunar eclipse, 12.44.
May	27	(Tue)	Saturn at opposition in Scorpius, magnitude 0·2.
July	10	(Thu)	Mars at opposition in Sagittarius, diameter 22″, magnitude –2.4.
Aug	11	(Mon)	Mercury at morning elongation (19°), favourable for northern observers.
Aug	12	(Tue)	Perseid meteor shower at maximum, 20.00. First Quarter Moon will soon set.
Aug	27	(Wed)	Venus at evening elongation.
Sep	10	(Wed)	Jupiter at opposition on the border of Aquarius and Pisces, magnitude –2·4.
Oct	3	(Fri)	Partial solar eclipse visible from Europe and the USA.
Oct	17	(Fri)	Total lunar eclipse, 19.19.
Oct	21	(Tue)	Mercury at evening elongation (24°), favourable for southern observers.
Nov	5	(Wed)	Venus passes inferior conjunction to become a morning object.
Nov	30	(Sun)	Mercury at morning elongation (20°), favourable for northern observers.
Dec	13	(Sat)	Geminid meteor shower at maximum, 24.00 (midnight). Full Moon will interfere.
Dec	23	(Tue)	Algol at minimum, 22.50.

Key Dates in Astronomical History

BC

c.3000 Babylonian astronomical records begin.

c.1000 Chinese astronomical records begin.

c.280 Aristarchus suggests that the Earth orbits the Sun.

c.270 Eratosthenes makes an accurate estimate of the size of the Earth.

c.130 Hipparchus draws up the first star catalogue.

AD

c.140 Ptolemy's *Almagest* written; his Earth-centred universe accepted.

903 Star positions measured by Al-Sufi.

1054 Supernova in Taurus recorded by Chinese astronomers.

1433 Ulugh Beigh's star catalogue compiled.

1543 Nicolaus Copernicus proposes a Sun-centred system.

1572 Supernova in Cassiopeia observed by Tycho Brahe.

1600 Johannes Kepler starts analysing Tycho's planetary observations and derives his three laws of planetary motion (1609-18).

1608 The refracting telescope invented by Hans Lippershey.

1609 Galileo and others make the first telescopic observations.

1631 Transit of Mercury across the Sun, predicted by Kepler, is observed by Gassendi.

1638 Holwarda discovers the famous variable star Mira Ceti.

1647 One of the first lunar maps is drawn by Hevelius.

1668 Isaac Newton constructs the first reflecting telescope.

▲ **Astronomers** in Istanbul Observatory in the Middle Ages. Notice the astronomical instruments they used.

1687 Newton's *Principia,* containing his theory of gravitation is published.

1705 Edmond Halley predicts that the comet last seen in 1682 will return again in 1758.

1725 The first 'modern' star catalogue, based on Flamsteed's observations, published.

1758 John Dollond manufactures the first successful achromatic object-glass. Halley's comet makes its predicted return.

1761 The first transit of Venus across the Sun observed.

1781 William Herschel discovers the new planet Uranus, and Charles Messier publishes his catalogue of star clusters and nebulae.

1801 Giuseppe Piazzi discovers the first asteroid, Ceres.

1838 Friedrich Wilhelm Bessel makes the first interstellar distance measurement, of 61 Cygni.

1840 The first astronomical photograph, of the Moon, taken by John William Draper.

1843 The sunspot cycle announced by Samuel Heinrich Schwabe.

1846 Neptune discovered as a result of predictions by John Couch Adams and Urbain Leverrier.

1859 Gustav Kirchhoff proves that the elements in a hot body imprint characteristic lines in its spectrum.

1870-1900 Great developments in astronomical photography and spectrum analysis. In 1891 George Ellery Hale invents spectroheliograph for photographing the Sun at a single wavelength.

1908 The Hertzsprung-Russell diagram introduces the idea of giant and dwarf stars. The first of the 'giant' reflecting telescopes, the 1·5-metre at Mount Wilson, begins work.

1912 The Cepheid period-luminosity law announced by Henrietta Leavitt.

1920 The red-shift first noticed in distant galaxies.

1923 Edwin Powell Hubble makes the first good measurement of the distance of M31 in Andromeda.

1930 Clyde Tombaugh discovers Pluto.

1930-1960 Many investigations into stellar energy production and the evolution of stars.

1937 The first radio waves from space detected by Grote Reber.

1948 The Mount Palomar 5-metre reflector completed.

1955 The Jodrell Bank 76-metre radio telescope completed.

1963 The great distances of quasars first established. Background radiation pervading space discovered.

1967 Pulsars discovered.

1977 The rings of Uranus discovered.

1981 The most remote known galaxies identified, at a distance of about 10,000 million light-years.

◄ **Isaac Newton's** reflector, the first of its kind.

Key Dates in Space Exploration

1804 The first high-altitude ascent made by Gay-Lussac and Biot in a hot-air balloon, reaching a height of seven kilometres.

1896 Unmanned balloons, launched by Teisserenc de Bort, analyse the atmosphere at heights of up to 15 kilometres.

1903 Ziolkovsky proposes a rocket-propelled spacecraft.

1919 Goddard publishes a monograph on rocket propulsion.

1923 Oberth's book *The Rocket into Interplanetary Space* published.

1926 Goddard launches the first liquid-fuelled rocket.

1942 Experiments with the V2 rocket at Peenemunde achieve heights of 180 kilometres.

1949 The first two-stage rocket, the WAC-Corporal, achieves a height of 400 kilometres.

1950 Cape Canaveral first used for rocket experiments.

1955 The USA announces its intention of launching space satellites.

1957 The world's first satellite, *Sputnik* 1, launched by the USSR on October 4.

1958 *Explorer* 1, the first satellite launched by the USA, discovers radiation belts around the Earth.

1959 Three lunar probes launched by the USSR: *Luna* 2 photographs the far side; *Luna* 3 hits the surface.

1961 The first manned orbital flight, by USSR astronaut Gagarin in *Vostok*.

▼ **One of two**
Voyager spacecraft that passed Saturn in 1980 and 1981. One is now heading for Uranus, the other for Neptune.

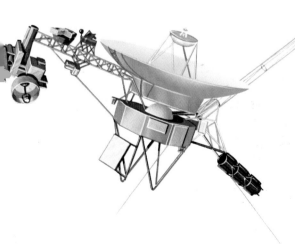

1962 The first successful inter-planetary probe, *Mariner 2,* sends back information about Venus.

1964 The first close-up pictures of the Moon obtained by *Ranger* 7 (USA).

1965 *Mariner* 4 (USA) passes Mars and transmits pictures and information.

1966 *Venera* 3 (USSR) lands on Venus, the first spacecraft to reach another planet. *Luna* 9 (USSR) makes the first soft landing on the Moon, followed by *Surveyor* 1 (USA). The first of the *Orbiter* mapping lunar satellites launched (USA).

1967 *Venera* 4 (USSR) soft-lands on Venus and sends back information.

1969 *Mariner* 6 *and* 7 (USA) pass Mars and send back pictures and information. *Apollo* 11 lands the first men on the Moon (July 20).

1970 The first automatic lunar probe, *Luna* 16 (USSR), returns a sample to Earth.

1971 *Mariner* 9 (USA) goes into orbit around Mars and sends back a great deal of information.

1973 The first flyby of Jupiter, by *Pioneer* 10, which becomes the first man-made artifact to escape from the solar system. *Skylab,* an orbiting astronomical laboratory, launched by the USA and visited by three different teams.

1974 *Mariner* 10 (USA) passes both Venus and Mercury. *Salyut* 3 *and* 4 (USSR) link up to form orbiting observatory. *Pioneer* 11 (USA) passes Jupiter and heads for Saturn.

1976 The first successful landings on Mars by the two *Viking* craft (USA).

1978 *Pioneer Venus* project (USA) puts two craft into orbit around Venus and lands four surface probes.

1979 *Voyagers* 1 *and* 2 (USA) pass Jupiter. *Pioneer* 11 passes Saturn successfully after a six-year journey.

1980 *Voyager* 1 makes successful flyby of Saturn, and heads for outer space.

1981 The first Space Shuttle, *Columbia* (USA), completes two flights. *Voyager* 2 passes Saturn and heads for Uranus, to reach it in 1986.

1982 *Columbia* (USA) completes third flight. USSR sends probes to Venus.

◄ **A lunarnaut** from *Apollo* 16 collects rock and soil samples.

Societies

If you want to find out more about astronomy, the best way of going about it is to join a local society. Societies are always on the lookout for new members, particularly for enthusiasts who want to learn

You will have the chance of looking through powerful telescopes, of borrowing books from the library (most societies have one), and of getting advice.

This is the most important benefit of all. You will probably find someone who has done astrophotography, someone else who has observed variable stars, and so on . . and you will learn far more by asking them questions than you will from books.

Ask at the local library or the civic centre if there is a society in your area, or else write to the **Federation of Astronomical Societies,** 1 Valletort Cottages, Millbridge, Plymouth, Devon PL1 5PU.

The Junior Astronomical Society (58 Vaughan Gardens, Ilford, Essex 1GI 3PD issues a quarterly journal, *Popular Astronomy,* which includes predictions of occultations, variable stars, and planet movements. The Society – for beginners of all ages – holds meetings and residential courses.

The senior amateur society in the UK is the **British Astronomical Association** (Burlington House, Piccadilly, London w1v 0NL), which was founded in 1890 and has a tremendous record of observational work.

Useful Books

Books to Read
The following books contain a great deal of useful information for the practical amateur:

Yearbooks and Atlases
The BAA Handbook, obtainable from the British Astronomical Association (address above).

Yearbook of Astronomy edited by P A Moore (Sidgwick & Jackson, annually).

Norton's Star Atlas and Reference Handbook (Gall & Inglis, 1978) is the best star atlas.

Guidebooks
Astronomy with Binoculars by James Muirden (Faber & Faber, 1976)

Field Guide to the Stars and Planets, by D H Menzel (Collins, 1978)

Practical Amateur Astronomy edited by P A Moore (Lutterworth, 1978).

Stars and Planets by Robin Kerrod (Kingfisher/Ward Lock, 1979)

The Practical Astronomer by Colin Ronan (Pan, 1980)

Glossary

Not all these words are used in the book but they are useful terms and ones that you may come across as an amateur astronomer.

Absorption spectrum Spectrum crossed by dark lines due to light absorbed by intervening cool gas.

Achromatic lens Used in a refracting telescope to give a colour-free image. A non-achromatic lens produces false colour fringes around objects in the sky.

Aerolite A meteorite consisting mainly of stony material.

Airglow Faint auroral luminosity of the night sky.

Albedo The ratio of light reflected to light received.

Altitude The height of an object above the horizon, measured in degrees (°).

Aphelion The point on the orbit of a planetary body that is farthest from the Sun.

Apogee The point on the orbit of a satellite that is farthest from the planet.

Astronomical Unit The mean distance of the Earth from the Sun (149,597,870 kilometres).

Binary Star A pair of stars orbiting around each other.

Chromosphere The inner atmosphere of the Sun.

Coma The head of a comet.

Comes The faint companion of a double star.

Conjunction Occurs when a planet, the Earth and the Sun are in line. In the case of Mercury and Venus which can come between the Earth and the Sun, they are said to be at inferior conjunction on the near side of the Sun and at superior conjunction on the far side.

Corona The Sun's outer atmosphere.

Declination A co-ordinate for finding objects in the sky. It is the equivalent of latitude on the Earth's surface.

Dichotomy The half-phase of the Moon or a planet.

Doppler shift The change of wavelength of sound or light waves due to motion between the source and the observer.

Eccentricity The word used to describe the difference between an ellipse and a circle. A very eccentric ellipse is a long thin loop.

Ecliptic is the path followed by the Sun around the year.

Ellipse The oval path traced by the planets and many comets.

Elongation The angle of Mercury or Venus from the Sun at any given time.

Emission spectrum Spectrum of bright lines, caused by luminous gas.

Eyepiece The lens or group of lenses in a telescope against which the eye is placed. It magnifies the image made by the object glass or mirror.

Fireball A meteor brighter than the planet Venus.

Focal length The distance between a lens or mirror and the image it forms of a remote object.

Fraunhöfer lines The absorption lines in the solar spectrum.

Galaxy A star system. The Galaxy refers to the galaxy to which our Sun belongs.

Gegenschein A very faint permanent glow in the sky opposite the Sun.

Granulation The fine mottled texture of the Sun's surface.

Hour angle The time in sidereal hours between a celestial object's present position and its meridian transit.

Ionosphere The upper layer of the Earth's atmosphere (above about 70 kilometres) where most atoms have either lost or gained electrons, and are therefore ionised.

Light-year Distance travelled by light in a year (9,460,700,000, 000 kilometres).

Limb The edge of the Sun, Moon or a planet as it is seen in the sky.

Luminosity A measure of light produced by a star.

Lunation The interval from one New Moon to the next.

Magnitude A scale that measures the brightness of a star.

Magnetosphere The shell of charged particles held around the Earth by its magnetic field.

Meridian An imaginary line crossing the sky and passing through zenith and the North and South celestial poles.

Midnight Sun The presence of the Sun above the horizon at midnight in high latitudes, when its distance from the celestial pole is smaller than the pole's altitude.

Node The point at which the orbit of the Moon or a planet crosses the plane of the Earth's orbit (ecliptic).

Occultation The covering of a celestial body by the Moon or a planet.

Opposition The instant when a planet is opposite in the sky to the Sun.

Parallax The slight shift in the position of a nearby star when viewed from opposite sides of the Earth's orbit.

Periastron The closest approach of two stars in a binary system as seen from the Earth.

Perigee The point on the orbit of a satellite that is closest to the planet.

Perihelion The point on the orbit of a planetary body that is closest to the Sun.

Perturbation The departure of a body from its true course due to the gravitational pull of another body.

Photosphere The visible surface of the Sun.

Prominence Eruptions of gas from the surface of the Sun.

Proper motion Drift of a star across the celestial sphere due to its own motion.

Radial velocity Motion of a star or galaxy towards or away from the observer.

Red-shift The way in which a star's colour reddens when it moves rapidly away from the observer. An approaching object becomes bluer, showing blue shift.

Retrograde motion Motion in the opposite sense from that followed by the planets in their orbits (anti-clockwise as seen from the north).

Saros The interval of 18 years 10·3 days, after which the Sun, Moon and Earth are in almost exactly the same relative position, and eclipses will recur.

Sidereal day The time taken by the Earth to rotate once as measured by a star: 23 hours, 56 minutes and 4 seconds long.

Siderite A meteorite consisting mainly of iron.

Solar day The time taken by the Earth to rotate once as measured by the Sun: 24 hours.

Solar wind The continuous outflow of atomic particles from the Sun.

Spectroheliograph An instrument for photographing the Sun in the light emitted by a single element.

Stratosphere The layer of calm, cold air lying between 15 and 40 kilometres above the Earth's surface.

Tektites Small glassy bodies, possibly of meteoric origin, found in a few regions of the Earth's surface.

Transit The passage of a smaller body across the disc of a larger one.

Troposphere The lower region of the Earth's atmosphere, extending to a height of about 15 kilometres.

Wavelength: The distance between pulses of radiation.

Wolf-Rayet stars Very hot stars with luminous atmospheres.

Zenithal Hourly Rate (ZHR) The number of meteors per hour that would be seen if the radiant were at the zenith.

Zodiacal Band An excessively faint band of light extending around the ecliptic, caused by interplanetary particles reflecting sunlight.

Index

ACKNOWLEDGEMENTS

Photographs: Ron Arbour 10, 70, 167 *top;* Arizona Meteor Laboratory 143 *right;* Lynne Bryant 32; California Institute of Technology 11
top, 42, 56, 58 *left and right,* 59, 61, 97, 154 *top,* 155; John Clarke
106/107; Dr Leo Connolly cover *top left;* P. Crabtree 83, 88, 166, 167
right and left; A. P. Dowdell 143 *left;* Hale Observatories 8, 57, 137
bottom; Alan W. Heath 105; Kitt Peak 18 *right;* Saul Levy 144; Lund
Observatory 152/3; Robin Kerrod back cover *bottom right*, 6, 18 *left,*
20; James Muirden front cover *bottom right*, 13, 18 *top*, 35, 38, 81, 87,
90, 95, 164, 165; NASA 7, 118, 119, 132, 137 *bottom*, 182, back cover
top left; New Mexico State University Observatory 125; Ann Ronan
147; Royal Astronomical Society 62, 64, 91, 100/1, 123, 136, 137 *top,*
140, 146, 154 *bottom*, 163; Royal Observatory, Edinburgh 60, 92;
Science Museum London 23, 180; Space Frontiers 11 *bottom,* 34
bottom, 63 *bottom*, 78, 127, 131, 139, 157, 155 *top*, back cover *top
right;* Science Photo Library 55, 148 (Jack Finch). front cover *top
right;* United States Naval Observatory 63 *top;* Yerkes Observatory 14
left; Zefa 41, 98, 145 (Photri), 161, back cover *bottom left.*

Picture Research: Jackie Cookson.